# Evolutionary Optimization of Material Removal Processes

This text comprehensively focuses on the concepts, implementation, and application of evolutionary algorithms for predicting, modeling, and optimizing the various material removal processes from their origin to the current advancements. This one-of-a-kind book encapsulates all the features related to the application and implementation of evolutionary algorithms for the purpose of predicting and optimizing the process characteristics of different machining methods and their allied processes that will provide comprehensive information. It broadly explains the concepts of employing evolutionary algorithm-based optimization in a broad domain of various material removal processes. Therefore, this book will enable prospective readers to take full advantage of recent findings and advancements in the fields of traditional, advanced, micro, and hybrid machining, among others. Moreover, the simplicity of its writing will keep readers engaged throughout and make it easier for them to understand the advanced topics.

This book

- Offers a step-by-step guide to implement evolutionary algorithms for the overall optimization of conventional and contemporary machining processes
- Provides in-depth analysis of various material removal processes through evolutionary optimization
- Details an overview of different evolutionary optimization techniques
- Explores advanced processing of various engineering materials-based case studies

It further discusses different nature-inspired algorithms-based modeling, prediction, and modeling of machining responses in attempting advanced machining of the latest materials and related engineering problems along with case studies and practical examples. It will be an ideal reference text for graduate students and academic researchers working in the fields of mechanical engineering, aerospace engineering, industrial engineering, manufacturing engineering, and materials science.

# Evolutionary Optimization of Material Removal Processes

Edited by
Ravi Pratap Singh, Narendra Kumar,
Ravinder Kataria, and Pulak Mohan Pandey

CRC Press
Taylor & Francis Group
Boca Raton London New York

CRC Press is an imprint of the
Taylor & Francis Group, an **informa** business

First edition published 2023
by CRC Press
6000 Broken Sound Parkway NW, Suite 300, Boca Raton, FL 33487-2742

and by CRC Press
4 Park Square, Milton Park, Abingdon, Oxon, OX14 4RN

*CRC Press is an imprint of Taylor & Francis Group, LLC*



*Library of Congress Cataloging-in-Publication Data*
Names: Singh, Ravi Pratap, editor.
Title: Evolutionary optimization of material removal processes / edited by Ravi Pratap Singh, Narendra Kumar, Ravinder Kataria, and Pulak Mohan Pandey.
Description: First edition. | Boca Raton : CRC Press, 2023. |
Includes bibliographical references and index.
Identifiers: LCCN 2022035921 (print) | LCCN 2022035922 (ebook) |
ISBN 9781032136516 (hardback) | ISBN 9781032192703 (paperback) |
ISBN 9781003258421 (ebook)
Subjects: LCSH: Machining--Data processing. | Cutting--Data processing. |
Evolutionary computation. | Mathematical optimization.
Classification: LCC TJ1185 .E96 2023 (print) | LCC TJ1185 (ebook) |
DDC 671.3/50285--dc23/eng/20221012
LC record available at https://lccn.loc.gov/2022035921
LC ebook record available at https://lccn.loc.gov/2022035922

ISBN: 978-1-032-13651-6 (hbk)
ISBN: 978-1-032-19270-3 (pbk)
ISBN: 978-1-003-25842-1 (ebk)

DOI: 10.1201/9781003258421

Typeset in Sabon
by MPS Limited, Dehradun

# Contents

# Acknowledgments

This book volume focuses in detail on the concepts, implementation, and application of evolutionary algorithms for predicting, modeling, and optimizing the various material removal processes from their origin to the current advancements. We owe many thanks to all the people who helped and supported us for the completion of this book project on time. We, the editors of this book, would like to express our gratitude towards all the esteemed authors for contributing their valuable research articles for this book. We would also like to acknowledge the reviewers for their painstaking and time-consuming effort in reviewing the chapters and providing their thorough evaluations for improving the quality.

We would further would like to express our sincere gratitude towards the entire team of CRC Press, Taylor & Francis Group who have given us time and space to write this book, in particular Gauravjeet Singh Reen, Senior Commissioning Editor, CRC Press for his continuous support and valuable guidance in completing this book proposal with this success. We would also like to express our sincere gratitude towards, Prof V K Jain, IIT Kanpur; Prof Pradeep Kumar, IIT Roorkee; and Prof Sandeep Singhal, NIT Kurukshetra, for their motivation and valuable inputs from time to time.

Lastly, we would like to thank to our families for believing in us and encouraging us to be the best we can be, and our friends, colleagues, and scholars for always being there with a helping hand and a good laugh.

# Preface

The recent advancements in the domain of material removal processes have been emerging for performing superior and classical performances while processing the newest and latest engineering materials. However, to deal with real-life industrial challenges, where it is required to predict and optimize the multiple/contradictory machining responses concurrently, some capable smart-algorithms-based optimization is needed. Evolutionary algorithms-based prediction and optimization often performs well by approaching solutions to all types of industrial issues. Therefore, the nature-inspired or bio-algorithms approaches, such as evolutionary algorithms, are highly significant areas considering their wide range of applications especially in the domain of material removal and allied processes. Hence, this book focuses on features related with the application and implementation of evolutionary or nature-inspired algorithms for the prediction and optimization of process responses in various material removal processes and allied methods. This volume contains 12 chapters and elaborates on the ongoing research in the domain of electric discharge machining, abrasive jet machining, turning operation, wire-cut EDM, hybrid machining, electrochemical machining, micro-electrochemical machining, high-speed drilling, CNC finish-turning, micro-machining, face turning, milling, etc. through statistical analysis, evolutionary modeling, and optimization and will be summarized on a common platform. The editorial team is comprised of renowned researchers and academicians in the field of traditional, advanced, micro, and hybrid machining. Their contributions in the book will enable the prospective readers to take the full benefits of recent findings and advancements in this field of material removal processes and allied methods.

This book volume focuses in detail on the concepts, implementation, and application of evolutionary algorithms for predicting, modeling, and optimizing the various material removal processes from their origin to the current advancements. This one-of-a-kind book encapsulates all the features related to the application and implementation of evolutionary algorithms for the purpose of predicting and optimizing the process characteristics of different machining methods and their allied processes that will provide comprehensive information. The idea is to author a book that is suitable for all types of aspirants of the domain of interest. This book broadly explains the concepts of employing evolutionary algorithm-based optimization in a broad domain of various material removal processes. Therefore, this book will enable the prospective readers to take full benefits from the recent findings and advancements in the fields of traditional, advanced, micro, hybrid machining, etc. Moreover, the simplicity of writing will keep the readers engaged throughout and make it easier for them to understand the advanced topics.

**Editors**

# Editors

**Dr. Ravi Pratap Singh** is as an assistant professor in the Dcpartment of Industrial and Production Engineering at the Dr. B. R. Ambedkar National Institute of Technology, Jalandhar, Punjab, India. Prior to this, he served in the Department of Technical Education, Uttar Pradesh, India, for about two years. He earned a doctor of philosophy (PhD) degree in the area of rotary ultrasonic machining of advanced engineering materials, and an MTech in mechanical engineering (industrial and production engineering) from the Department of Mechanical Engineering, National Institute of Technology, Kurukshetra (Haryana), India. He is a life member of the Indian Institution of Industrial Engineering (IIIE), Mumbai, India, and SCIence and Engineering Institute (SCIEI), Log Angeles, USA. Dr. Singh is also a senior member of the Indian Society of Mechanical Engineers (ISME), Chennai, India.

He has published more than 100 research articles throughout several SCI/Scopus indexed journals, including international/national-level conferences. He has also been engaged in the editing and review of several SCI/Scopus indexed journals for the last 9 to 10 years. He is also a guest editor of the *World Journal of Engineering* (Emerald Publications), *International Journal of Sustainable Materials and Structural Systems*, and *International Journal of Six Sigma and Competitive Advantage* (Inderscience Publisher). Dr. Singh has organized several national and international conferences at NIT Jalandhar as a convener/organizing secretary/session coordinator, among others.

Besides research excellence, he has received the Young Scientist in Mechanical Engineering in VIRA-2019 awards, the Young Faculty in Engineering (Major Area: Mechanical Engineering) in VIFA-2019 awards, and the Best Researcher Award in the 4th International Scientist Awards on Engineering, Science, and Medicine in January 2020 in New Delhi, India. He has delivered several expert talks/lectures at different levels throughout the country. He has also been nominated as a general chair reviewer, technical program committee, and national advisory committee, among others, during various national/international events held in China, Norway, and India. In addition to this, he is also serving as an ambassador of Bentham Open, Sharjah, UAE, as well as a member of the visor academic committee in Singapore. Dr. Singh has also been ranked among top 2 percent of scientists in the world as per a survey conducted by Stanford University in 2021.

**Dr. Narendra Kumar** is an assistant professor in the Industrial and Production Engineering Department at Dr. BR Ambedkar National Institute of Technology, Jalandhar, Punjab. He earned his PhD in mechanical engineering from Indian Institute of Information Technology, Design and Manufacturing, Jabalpur. His research work was based on the development and performance evaluation of pellet-based additive manufacturing process for flexible parts. His thesis project was part of a DST-sponsored project "Hybrid Additive-Subtractive Manufacturing System using CNC Machining Center." Prior to joining NIT Jalandhar, he worked for the Bajaj Institute of Technology, Wardha. He also worked on a DST/AMT-sponsored project titled "Development of Metal-Based Deposition System Using Induction Heating Method" as a research associate. His research interests are broadly related to additive manufacturing, machining methods, and material development. He has published more than 18 research papers in international journals and conference proceedings of high repute.

**Dr. Ravinder Kataria** is an assistant professor at National Institute of Fashion Technology (NIFT), Srinagar, Jammu & Kashmir, India. He earned a doctor of philosophy (PhD) degree in the area of ultrasonic machining of advanced engineering materials and a MTech in mechanical engineering (industrial and production dngineering) from the Department of Mechanical Engineering, National Institute of Technology, Kurukshetra (Haryana), India. He is a life member of SCIence and Engineering Institute (SCIEI), Log Angeles, USA. He has published more than 45 research articles throughout the several SCI/Scopus indexed journals, including national/international-level conferences. His broad areas of research are as follows: conventional and advanced manufacturing processes, advanced engineering materials, and micro-structure analysis. He is also an associate editor of *World Journal of Engineering* (Emerald Publications).

**Professor Pulak Mohan Pandey** completed his BTech degree from H.B.T.I. Kanpur in 1993, securing first position and earned a master's degree from IIT Kanpur in 1995 in manufacturing science specialization. He served at H.B.T.I. Kanpur as a faculty member for approximately 8 years and also earned a PhD in the area of additive manufacturing/3D printing from IIT Kanpur in 2003. He joined IIT Delhi as a faculty member in

2004 and is presently serving as professor. In IIT Delhi, Dr. Pandey diversified his research areas in the field of micro and nano finishing and micro-deposition and continued working in the area of 3D printing. He has supervised 32 PhD students and more than 36 MTech theses in the last 10 years and also filed 21 Indian patent applications. He has approximately 185 international journal papers and 45 international/national refereed conference papers to his credit. These papers have been cited more than 6,032 times with an h-index of 38. He received the Highly Commended Paper Award by *Rapid Prototyping Journal* for the paper "Fabrication of three dimensional open porous regular structure of PA 2200 for enhanced strength of scaffold using selective laser sintering" published in 2017. Many of his BTech- and MTech-supervised projects have been awarded by IIT Delhi. He is a recipient of the Outstanding Young Faculty Fellowship (IIT Delhi) sponsored by Kusuma Trust, and the Gibraltar and J.M. Mahajan outstanding teacher award of IIT Delhi. His students have won the GYTI (Gandhian Young Technological Innovation Award) in 2013, 2015, 2017, 2018, and 2020.

# Contributors

**Khaleel Ahmed**
Department of Mechanical Engineering
IMPACT college of Engineering & Applied Sciences
Bangalore, Karnataka, India

**Sachin P. Ambade**
Department of Mechanical Engineering
Yeshwantrao Chavan College of Engineering
Nagpur, Maharashtra, India

**Girija Nandan Arka**
Department of Production and Industrial Engineering
National Institute of Technology (NIT)
Jamshedpur, Jharkhand, India

**Jalumedi Babu**
Department of Mechanical Engineering
IMPACT college of Engineering & Applied Sciences
Bangalore, Karnataka, India

**Dilkush Bairwa**
Department of Industrial and Production Engineering
Dr. B R Ambedkar National Institute of Technology
Jalandhar, Punjab, India

**Dr Ravi Butola**
Department of Mechanical Engineering
Delhi Technological University
Delhi, India

**Shailendra Chauhan**
Department of Industrial and Production Engineering
Dr. B R Ambedkar National Institute of Technology
Jalandhar, Punjab, India

**Dr. Komal Ghanshyambhai Dave**
L D Engineering College
Ahmedabad, Gujarat, India

**J. Paulo Davim**
Department of Mechanical Engineering
University of Aveiro Campus
Santiago, Aveiro, Portugal

**Nikolaos A. Fountas**
Laboratory of Manufacturing Processes and Machine Tools (LMProMaT)
Department of Mechanical Engineering Educators
School of Pedagogical and Technological Education (ASPETE)
Amarousion, Greece

**Santosh Gupta**
Department of Metallurgical and Material Engineering
Visvesvaraya National Institute of Technology (VNIT)
Nagpur, Maharashtra, India

**Mohd Javaid**
Department of Mechanical Engineering
Jamia Millia Islamia
New Delhi, India

**Dr Ravinder Kataria**
National Institute of Fashion Technology
Srinagar, Jammu & Kashmir, India

**John Kechagias**
Design and Manufacturing Laboratory (DML)
University of Thessaly
Karditsa, Greece

**C M Krishna**
Maulana Azad National Institute of Technology
Bhopal, Madhya Pradesh, India

**Dr Narendra Kumar**
Department of Industrial and Production Engineering
Dr. B R Ambedkar National Institute of Technology
Jalandhar, Punjab, India

**Pawan Kumar**
Department of Mechanical Engineering
National Institute of Technology
Kurukshetra, Haryana, India

**Surinder Kumar**
Department of Mechanical Engineering
National Institute of Technology
Kurukshetra, Haryana, India

**Dimitrios E. Manolakos**
School of Mechanical Engineering
National Technical University of Athens (NTUA)
Zografou, Greece

**Meenu**
Department of Mechanical Engineering
National Institute of Technology
Kurukshetra, Haryana, India

**Sachchida Nand**
Amity University
Gwalior, Madhya Pradesh, India

**Madhusudan Painuly**
Department of Industrial and Production Engineering
Dr. B R Ambedkar National Institute of Technology
Jalandhar, Punjab, India

**Ioannis Papantoniou**
School of Mechanical Engineering
National Technical University of Athens (NTUA)
Zografou, Greece

**Jignesh Girishbhai Parmar**
Gujarat Technological University
Ahmedabad, Gujarat, India

**Lijo Paul**
Department of Mechanical Engineering
St. Joseph's College of Engineering & Technology
Kottayam, Kerala, India

**Abyson Scaria**
Department of Mechanical Engineering
St. Joseph's College of Engineering & Technology
Kottayam, Kerala, India

**Sagar Shelare**
Department of Mechanical Engineering
Priyadarshini College of Engineering
Nagpur, Maharashtra, India

**Manvandra K Singh**
Amity University
Gwalior, Madhya Pradesh, India

**Dr Ravi Pratap Singh**
Department of Industrial and Production Engineering
Dr. B R Ambedkar National Institute of Technology
Jalandhar, Punjab, India

**Subhash Singh**
Department of Mechanical and Automation Engineering
Indira Gandhi Delhi Technical University for Women
New Delhi, India

**Dr Sandeep Singhal**
Department of Mechanical Engineering
National Institute of Technology
Kurukshetra, Haryana, India

**Chetan K. Tembhurkar**
Department of Mechanical Engineering
Priyadarshini College of Engineering
Nagpur, Maharashtra, India

**Nikolaos M. Vaxevanidis**
Laboratory of Manufacturing Processes and Machine Tools (LMProMaT)
Department of Mechanical Engineering Educators
School of Pedagogical and Technological Education (ASPETE)
Amarousion, Greece

# Introduction

Evolutionary algorithms-based prediction and optimization often perform well, resembling solutions to all types of industrial issues as they ideally do not make any supposition about the fundamental fitness landscape. Over a period of time, it is desired to make the existing manufacturing smarter and more sustainable by the proper implementation of numerous well-proven mathematical modeling and nature-inspired algorithm-based prediction and optimizations namely as; particle swarm optimization, ant bee colony, teacher-learner based optimization, jaya algorithm, genetic algorithm, artificial neural network, fuzzy-based multi-criterion decision-making approaches, etc. while attempting the tasks at the real-life level.

This book offers a detailed structure of evolutionary or nature-inspired optimization approaches broadly employed to model and optimize the various types of advanced and traditional machining processes, fabrication, and allied methods. The different statistical methods and evolutionary approaches applied for the purpose to optimize the single and/or multiple objectives for overall enhancement of process quality and improvement by making the advanced machining solutions more sustainable are covered in this book. Furthermore, the various recent and latest research and development works in the broad domain of material removal processes have also been reported by emphasizing real-life case study-based elaborations. In addition, the readers will also be able to develop these processes more sustainably and reliably by making an effective implementation of evolutionary approaches for modeling and optimization in the broad domain of machining of various engineering materials.

This book will further edify its readers about the current and latest research interests going on around the globe in the field of various material removal processes and its allied domains. The recent advancements in numerous machining practices offer several fruitful solutions while processing of the latest and difficult-to-machine materials used in different industrial sectors will also be focused on. The ongoing research in the broad domain of electric discharge machining, abrasive jet machining, turning operation, wire-cut EDM, hybrid machining, electrochemical machining, micro-electrochemical machining, high-speed drilling, CNC

DOI: 10.1201/9781003258421-1 

finish-turning, micro-machining, face turning, milling, etc. through evolutionary modeling and optimization will be summarized on a common platform. Furthermore, this book will also be highly valuable in order to learn ideas, knowledge, skill, and experience shared by valuable authors, researchers, and scientists in the domain of interest.

Editors

Chapter 1

# Experimental Investigation of Surface Roughness for Turning of UD-GFRP Composite Using PSO, GSA, and PSOGSA Techniques

*Meenu*[1] *and Surinder Kumar*[2]

[1]Professor, Department of Mechanical Engineering, National Institute of Technology, Kurukshetra, Haryana, India

[2]Assistant Professor, Department of Mechanical Engineering, National Institute of Technology, Kurukshetra, Haryana, India

**CONTENTS**



## 1.1 INTRODUCTION

The heterogeneous materials (composites) are a shared mixture of two or extra micro-constituents that are different in chemical composition and substantial form and are not soluble in all others. Fiber glass composites are economic materials that replace many of the materials used in

DOI: 10.1201/9781003258421-2

industries. The FRPs are difficult to machine due to the arrangement of fibers. Discontinuity is established in the fiber due to how machining affects the performance of the part (Abrate and Walton, 1992). Konig et al. (1985) found that protruding fiber tips in composites can give incorrect outcomes.

Glass fiber–reinforced plastic is extensively used in applications such as buildings, piping, and robots. The advantages are high fracture toughness, thermal resistance, and corrosion. There are many methods for machining composite materials, such as ultrasonic machining, laser cutting, and EDM. The lower rate of production is the disadvantage of non-conventional machining. So, conventional machining processes for machining composite materials still find a wider acceptance. The mechanism of material removal rate in composite material is different from metals. The special consideration must be given to the wear resistance of the tool while machining fiber-reinforced materials. Hence, suitable cutting tool materials used are carbide and diamond tools (Davim and Reis, 2004).

## 1.2 LITERATURE REVIEW

Santhanakrishnan et al. (1989) studied turning of Kevlar fiber, GFRP, and CFRP composite using a K20 carbide and HSS tool. The performance of a K20 carbide tool for machining fiber-reinforced plastic composites was better than others. Palanikumar et al. (2006) optimized particular cutting pressure, tool wear, and surface smoothness throughout the turning of glass FRP composites with a carbide tool. Fiber orientation of workpiece, depth, feed, speed, and machining time were considered for the minimization of surface roughness. The machining performance obtained was superior when more parameters and levels were considered.

Isik (2008) obtained the results of the machining of UD-GFRP and the most favorable cutting parameters were recommended to acquire a desired surface quality. Surinder Kumar et al. (2013) studied the machinability of UD-GFRP via a K10 carbide cutting tool. Six parameters were considered in the MRR and to diminish surface roughness by ANOVA. Utility function was significantly affected by speed, DOC (depth of cut), and feed. Hussain et al. (2011) utilized Taguchi's L25 OA for experiments and fuzzy logic for modeling during the turning of GFRP using a K20 carbide tool. Parveen Raj et al. (2012) used mathematical modeling and ANN modeling to expect delaminating and surface roughness while end milling of the GFRP. The hand layup process was used to construct specimens of eight-layered UD-GFRP. The tools used were carbide K10 (solid) and titanium coated carbide end mill. Tool material, feed, speed, and depth of cut were the parameters for study. Farhad and Mahdi (2008) used simulated annealing to optimize the turning parameters while machining of cylindrical workpieces. The goal was to minimize the total cost. Seven constraints were used. Yang et al. (2009) used simulated annealing for parameter value selection for an electric discharge process by minimizing surface roughness.

Several techniques to resolve optimization issues found in the literature are gravitational search algorithm (Esmat et al. 2009), gray wolf algorithm (Mirjalili et al., 2014), simulated annealing algorithm (Suman and Kumar, 2006), etc. These are non-deterministic methods that are used to find optimum solution in complex problems, as it is difficult to explore whole domains of possible solutions in a reasonable time by conventional methods (Battle, 2008). These algorithms have been useful to a large range of applications.

In this research work, PSO, GSA, and PSOGSA are compared to discover the best possible parameters for turning of a UD-GFRP composite. PSO is a universal optimization method that has been suggested by Eberhart and Kennedy based on the flock of animals and school of catch fish (Eberhart and Kennedy, 1995).

Esmat et al. (2009) used GSA to resolve optimization problems. There have been modifications in the algorithm of GSA since its introduction. GSA has beena solution to the diversity of applications. Various variants of GSA are available that enhance the original version. A number of customized versions of GSA introduced are nonstop, binary-valued, triple-valued, separate, restraint, multi-objective, and multimodal. In GSA, the gravitational force transfers information between different masses. The results obtained by GSA are better quality and comparable with RGA, PSO, and CFO. Esmat Rashedi et al. (2011) used GSA to put up a pass through a filter modeling and compared it with PSO and GA and the outcome showed that GSA performed better than PSO and GA. The performance in the exploration and exploitation was improved. However, in recent times, GSA has been criticized because it is considered that it is not truly based on the gravitational law. Debkalpa Goswami and Shankar Chakraborty (2015) applied the fireworks algorithm (FWA) and gravitational search algorithm on behalf of optimization of the USM process. It was noticed that the FWA provides the best possible results for the USM process.

Sajad A.R. and Shanthi B. (2020) solved the problem of mechanical design by chaotic gravitational search algorithm (CSGA). The CGSA gave the efficient performance in terms of optimization of design variable, convergence speed, handling of constraints, and minimization of cost function compared to other algorithms. Mirjalili et al. (2010) developed the latest hybrid population-based algorithm (PSOGSA) that integrates local search capability of GSA with social thinking of PSO to optimize the benchmark problems.

In this work, the influence of DOC, environment, speed, feed, nose radius, and rake angle on the surface irregularity during turning has been considered. The design of experiments by Taguchi is used to measure surface roughness. Only significant parameters are mathematically modeled. The objective function is the mathematical model to be optimized by GSA, PSO, and PSOGSA. PSOGSA overcame the problem of premature convergence of PSO (Gao et al., 2008). PSOGSA obtained the best cutting conditions and tool geometry.

## 1.3 EXPERIMENTAL PROCEDURE

The Taguchi $L_{18}$ orthogonal array (OA) with 17 ($2^1 * 3^7$) degree of freedom is used. The nose radius is kept at two levels and other variables are kept at three levels. To assign the process variables, a linear graph is used. The process parameter and their ranges and symbols are presented in Table 1.1. The 18 tests used all have six parameters assigned to columns 1 toward 6, as indicated in Table 1.2.

*Table 1.1* Input variables and their level

| *Input Parameters* | *Levels* | | |
|---|---|---|---|
| | *L1* | *L2* | *L3* |
| Nose radius (x1) | 0.4 mm | 0.8 mm | Nothing |
| Rake angle (x2) | −6° | 0° | +6° |
| Feed (x3) | 0.05 mm/rev | 0.1 mm/rev | 0.2 mm/rev |
| Speed (x4) | 55.42 m/min | 110.84 m/min | 159.66 m/min |
| Cutting environment (x5) | Dry | Wet | Cooled |
| DOC (x6) | 0.2 mm | 0.8 mm | 1.4 mm |

*Table 1.2* Layout for experiments using orthogonal array ($L_{18}$)

| *Expt. no.* | *x1* | *x2* | *x3* | *x4* | *x5* | *x6* | *nil* | *nil* |
|---|---|---|---|---|---|---|---|---|
| 1 | 1 | 1 | 1 | 1 | 1 | 1 | 1 | 1 |
| 2 | 1 | 1 | 2 | 2 | 2 | 2 | 2 | 2 |
| 3 | 1 | 1 | 3 | 3 | 3 | 3 | 3 | 3 |
| 4 | 1 | 2 | 1 | 1 | 2 | 2 | 3 | 3 |
| 5 | 1 | 2 | 2 | 2 | 3 | 3 | 1 | 1 |
| 6 | 1 | 2 | 3 | 3 | 1 | 1 | 2 | 2 |
| 7 | 1 | 3 | 1 | 2 | 1 | 3 | 2 | 3 |
| 8 | 1 | 3 | 2 | 3 | 2 | 1 | 3 | 1 |
| 9 | 1 | 3 | 3 | 1 | 3 | 2 | 1 | 2 |
| 10 | 2 | 1 | 1 | 3 | 3 | 2 | 2 | 1 |
| 11 | 2 | 1 | 2 | 1 | 1 | 3 | 3 | 2 |
| 12 | 2 | 1 | 3 | 2 | 2 | 1 | 1 | 3 |
| 13 | 2 | 2 | 1 | 2 | 3 | 1 | 3 | 2 |
| 14 | 2 | 2 | 2 | 3 | 1 | 2 | 1 | 3 |
| 15 | 2 | 2 | 3 | 1 | 2 | 3 | 2 | 1 |

*(Continued)*

*Table 1.2* (continued)

| *Expt. no.* | *x1* | *x2* | *x3* | *x4* | *x5* | *x6* | *nil* | *nil* |
|---|---|---|---|---|---|---|---|---|
| 16 | 2 | 3 | 1 | 3 | 2 | 3 | 1 | 2 |
| 17 | 2 | 3 | 2 | 1 | 3 | 1 | 2 | 3 |
| 18 | 2 | 3 | 3 | 2 | 1 | 2 | 3 | 1 |

*Figure 1.1* Specimen UD-GFRP rod.

Specimens with glass (75±5%) and epoxy resin (25±5%) are made, as shown in Figure 1.1. The details of UD-GFRP specimens are modulus of elasticity 320 N/mm$^2$, weight of rod 2.300 kg, shear strength 255 N/mm$^2$, length 840 mm, reinforcement UD (E' Glass Roving), diameter 42 mm, density 1.95–2.1 gm/cc, and water absorption 0.07%. The geometry of the cutting bit used is as follows: nose radius (0.4 mm, 0.8 mm), rake angle [−6°, 0°, +6°,] and front clearance 10°. The Tokyo Seimitsu Surf-com 130A type instrument is used to measure surface irregularity ($R_a$). The transverse length of 4 mm is used to set up a carry on a length of 0.8 mm.

## 1.4 METHODOLOGY

Experimental data is collected using the Taguchi approach. The second-order model is developed using experimental data. The objective function i.e. mathematical model is optimized using GSA, PSO, and PSOGSA approaches.

### 1.4.1 Taguchi Method

Taguchi methodology is an effective technique used to optimize output quality characteristics by various procedure parameters. It helps in

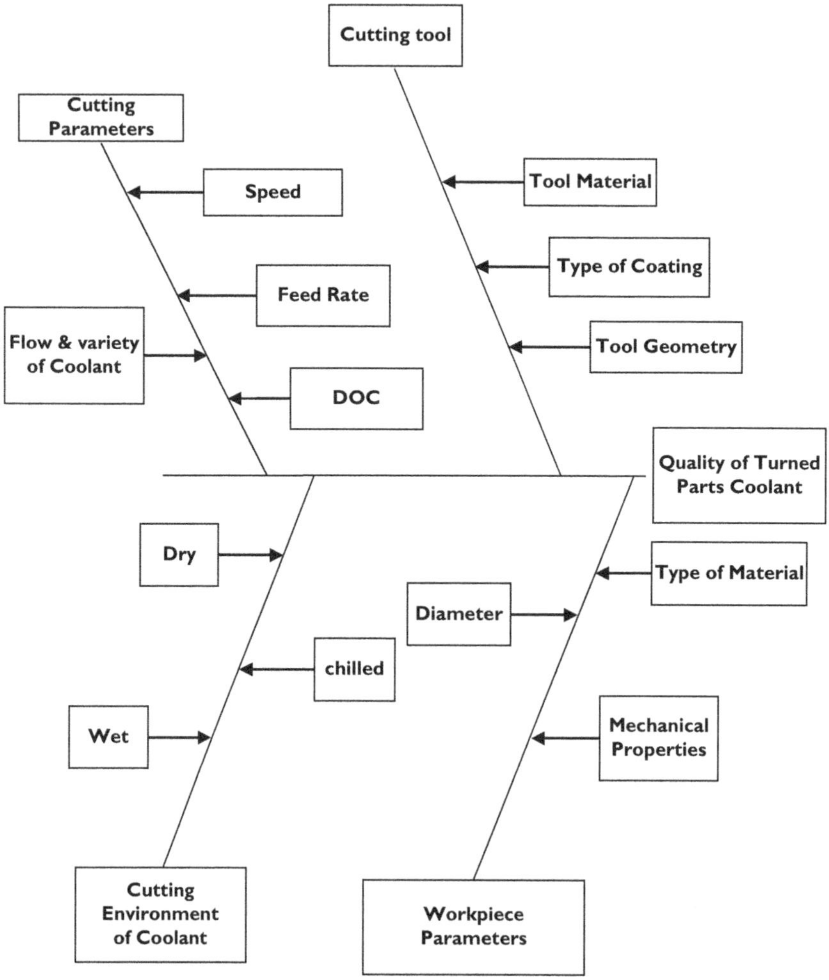


*Figure 1.2* Ishikawa diagram.

obtaining a good mixture of experiments to make experimental design easy and finely organized (Rose, 1996). In order to achieve this, an Ishikawa diagram is made, as shown in Figure 1.2.

## 1.4.2 Multiple Regression Methodology

(Douglas Montogomery et al., 2001) Mathematical models using single and multiple regressions analysis are developed for surface roughness, $R_a$. The regression equation correlates the dependent variables with the independent variables, such as DOC, cutting environment, feed, cutting speed, rake angle, and tool nose radius. The first-order model gave low predictability i.e. high prediction error for measured responses. So, a second-order model is used for regression analysis (Douglas Montogomery et al., 2001).

### 1.4.3 Gravitational Search Algorithm

GSA is a metaheuristic algorithm motivated by Newton's law of gravity and motion and is utilized to work out optimization issues. Gravity is one of the four essential connections in life (Schutz, 2003). Each element in the space attracts another particle with a force in inverse proportion to the square of distance between them and in proportion to M1*M2. The masses follow the law of gravity and motion, as given in Equations (1.1) and (1.2), respectively.

$$\mathrm{F} = \mathrm{G} * \frac{M1 * M2}{R^2} \tag{1.1}$$

$$\mathrm{a} = \frac{F}{M} \tag{1.2}$$

Where F: the gravitational force, G: gravitational constant, M1 and M2: the mass of the initial and next particle, and R: the span between the two masses. In GSA, each mass has four terms associated with it: active mass, passive mass, inertial mass, and position (Esmat Rashedi et al., 2009). The steps of GSA are as follows:

Step 1. **Initialization of particles**

The particles are initialized at random.

$$X_i = (x_i^1, \ldots x_i^d, \ldots, x_i^n), \text{ for i} = 1, 2, .., \mathrm{N}. \tag{1.3}$$

$x_i^d$ is the value of the $i^{th}$ particle for the d$^{th}$ variable, while n is the number of variables and N is the population size.

Step 2. **Fitness evaluation and best fitness computation**

The excellent and most undesirable solutions are found at each iteration.

**For Minimization:**

$$\mathrm{best(t)} = \min \mathrm{fit_j(t)} \text{ where } \mathrm{j} \varepsilon \{1, \ldots N\} \tag{1.4}$$

$$\mathrm{worst(t)} = \max \mathrm{fit_j(t)} \text{ where } \mathrm{j} \varepsilon \{1, \ldots N\} \tag{1.5}$$

**For Maximization:**

$$\mathrm{best(t)} = \max \mathrm{fit_j(t)} \text{ where } \mathrm{j} \varepsilon \{1, \ldots N\} \tag{1.6}$$

$$\text{worst}(t) = \min \text{fit}_j(t) \text{ where } j\varepsilon\{1, \ldots N\} \tag{1.7}$$

*fitj*(t) is the condition of the j$^{th}$ agent at iteration t, best(t) is the top fitness obtained at iteration t, worst(t) is the most unfavorable condition at iteration t.

Step 3. **Gravitational constant (G) computation** (Chatterjee, 2011):

$$G(t) = G_O e^{\left(-\frac{\alpha t}{T}\right)} \tag{1.8}$$

$G_O$ and α are initialized that reduces with the iterations. The iterations are specified by T.

Step 4. **Mass computation:**

Gravitational and inertia masses for every particle is considered as $M_{ai} = M_{pi} = M_{ii} = M_i i = 1, 2 \ldots \ldots, N$.

$$m_i(t) = \frac{Fit_i(t) - worst(t)}{best(t) - worst(t)} \tag{1.9}$$

$$M_i(t) = \frac{M_i(t)}{\sum_{j=1}^{N} M_i(t)} \tag{1.10}$$

$M_{pi}$ and $M_{ai}$ are the passive and active masses, while $M_{ii}$ is the inertia mass of the i$^{th}$ particle

Step 5. **Calculation of acceleration of particles:**

At $t^{th}$ iteration, acceleration of the $i^{th}$ agents is found as:

$$a_i^d(\text{t}) = \frac{F_i^d(t)}{M_{ii}(t)} \tag{1.11}$$

$Fi^d(t)$ = The total force acting on $i^{th}$ agent given by Equation (1.12).

$$Fi^d(\text{t}) = \Sigma \text{randj } F_{ij}^d(\text{t}) \; j \; \varepsilon \; Kbest, \; j \neq i \tag{1.12}$$

Kbest is the first K particles having the best fitness value, which decreases linearly with time. At the end, there is one particle that applies force to others, which is given by Equation (1.13).

$$F_{ij}^d(\text{t}) = G(\text{t}). \frac{M_{pi}(\text{t}) * M_{aj}(\text{t})}{R_{ij}(\text{t}) + \varepsilon}. (x_j^d(\text{t}) - x_i^d(\text{t})) \tag{1.13}$$

**Algorithm Steps:**

**1:** Create the population

**2:** Evaluate each individual.

**3:** Find new G(t), finest(t), most worst(t) and $M_i(t)$ for i = 1, 2,....N.

**4:** Estimation of the force in dissimilar directions.

**5:** Computation of increase in velocity.

**6:** Updation of the velocity and location.

**7:** Repetition of steps 2 to 6 until the stop criterion is reached.

**8:** End.

*Figure 1.3* The algorithm of GSA.

$F_{ij}^{d}(t)$ is the force acting on particle i from particle j in the $d^{th}$ direction at the $t^{th}$ iteration. $R_{ij}(t)$ is the Euclidian span between two particles, i and j, at the iteration t. G(t) is the gravitational constant at the $t^{th}$ iteration. ε is a very small constant.

Step 6. **Velocity and position of particles:**

Velocity and the location of the particle at the (t+1) iteration is found based on the subsequent equations:

$$v_i^d(t+1) = \text{randi} * v_i^d(t) + a_i^d(t) \tag{1.14}$$

$$x_i^d(t+1) = x_i^d(t) + v_i^{d_d}(t+1) \tag{1.15}$$

Repeat steps 2–6 until the iterations arrive at their upper bound. At the final iteration, the best fitness gives global fitness and the position of particle gives global solution. Figure 1.3 shows the algorithm of GSA.

### 1.4.4 Particle Swarm Optimization

Each particle in the swarm represents the candidate solution for the problem. In PSO, each particle changes its position and moves to the latest situation and memories of the top location encountered and the top position of neighbors to move towards the universal minimum. PSO is a computationally efficient, uncomplicated concept. PSO is a flexible method that improves the local exploitation and global abilities. The principle of the PSO algorithm is as follows (Esmin et al., 2005). The PSO consists of a swarm with N particles in a d-dimension (i.e. d variables). The location and speed of individuals are described as:

$$X_i = (x_i^1 \ldots\ldots\ldots x_i^{d,}) \text{ for } i = 1, 2, 3 \ldots N \tag{1.16}$$

**1:** Create the initial group having N particles at arbitrary positions.
**2:** Produce the starting velocity arbitrarily.
**3:** Detect the outstanding solution that has been achieved by the particle from 1 to t iterations and excellent solution obtained so far by any element of the population.
**4:** Modify the velocity and position.
**5:** If the stopping situation is fulfilled, stop.
Else, go away to Step 3.

*Figure 1.4* The algorithm of PSO.

$$V_i = (v_i^1 \ldots\ldots\ldots..v_i^d) \quad (1.17)$$

Equations (1.18–1.19) are used to get the updated velocity and position of the i$^{th}$ particle. The algorithm of PSO is shown in Figure 1.4.

$$v_i^j(t+1) = w * v_i^j(t) + c_{1*}rand * (p_{i,}^j(t) - x_i^j(t)) + c_{2*}rand * (g^j(t) - x_i^j(t) \quad (1.18)$$

Where $v_{i,}^j(t)$: Velocity of the $i^{th}$ particle in the j$^{th}$ dimension (variable) at iteration t

$x_i^j(t)$: The $i^{th}$ particle position in the j$^{th}$ dimension (variable) at iteration t

$c_1$, $c_2$: load factors

$p_{i,}^j(t)$: The $i^{th}$ particle finest position in the j$^{th}$ dimension (variable) at iteration t

$g^j(t)$: The top location achieved so long by any individual in the j$^{th}$ dimension at iteration t

$r_i$ and $r_2$: arbitrary numbers $\epsilon$ [0,1]

w: inertia weight

$$x_i^j(t+1) = x_i^j(t) + v_i^j(t) \quad (1.19)$$

## 1.4.5 Hybridized PSOGSA

In hybrid PSOGSA (Seyedali Mirjalili, 2010), the functionality of both are combined. In this, the local search capability of GSA is combined with the social thinking of PSO. These algorithms are combined and the equation for velocity and position is given by Equations (1.20–1.21).

Figure 1.5

$$v_i^j(t+1) = w * v_i^j(t) + c_1' * rand * ac_i^j(t) + c_2' * rand * (g^j(t) - x_i^j(t)) \quad (1.20)$$

Figure 1.5 shows the algorithm of PSOGSA.

Where $v_i^j(t)$: the speed of the i$^{th}$ particle in the j$^{th}$ dimension at iteration t

$c_1' and c_2'$: weighting factors.

**1:** Initially produce *N* particle at arbitrary.
**2:** Estimate of most excellent fitness, worst fitness, Gravitational constant, mass, acceleration and force using Equations 9-14
**3:** Locate the most excellent solution that has been achieved by the particle
**4:** Revise the velocity and location as given by Equation 21 and 22
**5:** If the stop situation is realised than, stop.
Otherwise, go to Step 3.

*Figure 1.5* The algorithm of PSOGSA.

w: weighting function.
rand: a arbitrary numeral between 0 and 1.
$g^j(t)$: best position achieved so long by particle i in the j$^{th}$ dimension at iteration t
$ac_i^j(t)$: the acceleration of individual in the j$^{th}$ dimension at iteration t
$g^j(t)$: the best solution in the j$^{th}$ dimension at iteration t obtained so far.

$$x_i^j(t+1) = x_i^j(t) + v_i^j(t) \tag{1.21}$$

## 1.5 RESULTS AND DISCUSSION

Table 1.3 shows the 18 trial conditions and surface roughness, $R_a$, for experimental parameter combination. The average ($R_a$) of 1.570 µm is achieved in experiment 17 at DOC of 0.2 mm, cooled cutting environment, speed of 55.42 m/min, feed of 0.1 mm/rev, rake angle of +6°, and nose radius of 0.8 mm. Combination of cooled environment, moderate feed, lowest speed, and DOC and largest rake angle and nose radius resulted in a better surface finish. The response table for surface irregularity ($R_a$) is shown in Table 1.4. Table 1.4 shows that the feeds that have the highest contribution (Δ = highest-lowest = 0.816) followed by speed (Δ = highest-lowest = 0.717) and DOC (Δ = highest-lowest = 0.508).

### 1.5.1 Analysis of Variance

From ANOVA Table 1.5, it is found that the feed is the most important factor that affects surface irregularity. The environment, rake angle, and nose radius are found to have little effect. The best possible combination is: DOC (0.2 mm), cutting environment (wet), speed (110.84 m/min.), feed (0.1 mm/rev.), rake angle (+6 degree), and nose radius (0.8 mm). It is obvious from ANOVA that the DOC, speed, and feed affect significantly the mean value as fit as the variant of the $R_a$. The percent contributions of DOC, speed, and feed rate are 10.584%, 21.595%, and 29.110%, respectively. From the ANOVA outcome, it is understood that DOC, speed, and feed have an important effect on $R_a$ and nose radius, tool rake angle, and surroundings have

*Table 1.3* Results of experimentation

| Expt. No. | Input Parameters | | | | | | Responses | | | |
|---|---|---|---|---|---|---|---|---|---|---|
| | | | | | | | Raw data | | | Average surface roughness |
| | Nose Radius (x1) | Rake angle (x2) | Feed (x3) | Speed (x4) | Cutting environment (x5) | Depth of cut (x6) | Surface Roughness ($R_a$) 1 2 3 | | | |
| | | | | | | | 1 | 2 | 3 | |
| 1 | 0.4 | −6° | 0.05 | 55.42 | Dry | 0.2 | 1.59 | 1.65 | 1.49 | 1.577 |
| 2 | 0.4 | −6° | 0.1 | 110.84 | Wet | 0.8 | 1.73 | 1.77 | 1.99 | 1.830 |
| 3 | 0.4 | −6° | 0.2 | 159.66 | Cooled | 1.4 | 2.77 | 4.12 | 5.13 | 4.00 |
| 4 | 0.4 | 0° | 0.05 | 55.42 | Wet | 0.8 | 2.20 | 2.18 | 2.04 | 2.140 |
| 5 | 0.4 | 0° | 0.1 | 110.84 | Cooled | 1.4 | 1.83 | 1.83 | 1.77 | 1.810 |
| 6 | 0.4 | 0° | 0.2 | 159.66 | Dry | 0.2 | 2.69 | 2.88 | 2.89 | 2.820 |
| 7 | 0.4 | +6° | 0.05 | 110.84 | Dry | 1.4 | 1.62 | 1.94 | 2.12 | 1.893 |
| 8 | 0.4 | +6° | 0.1 | 159.66 | Wet | 0.2 | 1.99 | 1.79 | 1.89 | 1.890 |
| 9 | 0.4 | +6° | 0.2 | 55.42 | Cooled | 0.8 | 2.58 | 2.94 | 2.10 | 2.540 |
| 10 | 0.8 | −6° | 0.05 | 159.66) | Cooled | 0.8 | 2.90 | 2.72 | 2.35 | 2.656 |
| 11 | 0.8 | −6° | 0.1 | 55.42 | Dry | 1.4 | 2.15 | 2.20 | 1.95 | 2.100 |
| 12 | 0.8 | −6° | 0.2 | 110.84 | Wet | 0.2 | 2.45 | 1.56 | 2.26 | 2.090 |
| 13 | 0.8 | 0° | 0.05 | 110.84 | Cooled | 0.2 | 1.77 | 1.55 | 1.89 | 1.736 |
| 14 | 0.8 | 0° | 0.1 | 159.66 | Dry | 0.8 | 3.05 | 2.41 | 2.51 | 2.656 |
| 15 | 0.8 | 0° | 0.2 | 55.42 | Wet | 1.4 | 2.61 | 1.87 | 3.38 | 2.620 |
| 16 | 0.8 | +6° | 0.05 | 159.66 | Wet | 1.4 | 2.26 | 2.69 | 1.96 | 2.303 |
| 17 | 0.8 | +6° | 0.1 | 55.42 | Cooled | 0.2 | 1.65 | 1.68 | 1.38 | 1.570 |
| 18 | 0.8 | +6° | 0.2 | 110.84 | Dry | 0.8 | 2.53 | 2.99 | 2.50 | 2.673 |
| | | | **Total** | | | | | | | **Overall Mean = 2.272** |

*Table 1.4* Response for surface roughness

| *Levels* | *(x1)* | *(x2)* | *(x3)* | *(x4)* | *(x5)* | *(x6)* |
|---|---|---|---|---|---|---|
| L1 | 2.279 | 2.377 | 2.051 | 2.091 | 2.287 | 1.947 |
| L2 | 2.267 | 2.297 | 1.976 | 2.006 | 2.146 | 2.416 |
| L3 | — | 2.145 | 2.792 | 2.722 | 2.387 | 2.456 |
| Delta | 0.011 | 0.232 | 0.816 | 0.717 | 0.241 | 0.508 |
| Rank | 6 | 5 | 1 | 2 | 4 | 3 |

*Table 1.5* ANOVA results for surface roughness

| *Source* | *Sum of square* | *Degree of freedom* | *Variance* | *F ratio* | *Prob.* | *Pure sum of square SS'* | *Percent contribution P (%)* |
|---|---|---|---|---|---|---|---|
| Nose radius of tool (x1) | 0.0017 | 1 | 0.0017 | error | 0.922 | — | — |
| Rake angle of tool (x2) | 0.4989 | 2 | 0.2495 | error | 0.245 | — | — |
| Feed rate (x3) | 7.3151 | 2 | 3.6575 | 21.30* | 0.000 | 6.972 | 29.110 |
| Speed (x4) | 5.5154 | 2 | 2.7577 | 16.06* | 0.000 | 5.172 | 21.595 |
| Cutting Environment (x5) | 0.5283 | 2 | 0.2641 | error | 0.227 | — | — |
| DOC (x6) | 2.8789 | 2 | 1.4394 | 8.38* | 0.001 | 2.535 | 10.584 |
| Total error | 23.9501 | 53 | | | 23.9501 | 100.00 | |
| (pooled) | 7.2119 | 42 | | | 0.1717 | 9.101 | 37.99 |

no consequence at a 95% confidence level. The DOC is found to be the least important parameter and feed rate as the most significant factor. The feed rate is the for the most part important factor that affects surface irregularity.

## 1.5.2 Multiple Regression Prediction Model

A multiple regression equation is modeled to find a relationship between process parameters and $R_a$ for several combination of factor stage in a particular range. So, a second-order model is used for regression analysis in Equation (1.22).

$$\begin{aligned} R_a = {} & 7.35 + 1.41\, x_1 + (-6.61)\, x_2 + 0.268\, x_3 + 0.312\, x_1x_2 + 0.133\, x_1x_3 \\ & + (-0.069)\, x_2x_3 + 0.898\, {x_1}^2 + 1.82\, {x_2}^2 + (-0.229)\, {x_3}^2 \end{aligned} \tag{1.22}$$

Where $R_a$ and different predictors are log values. The predicted and measured experiential values for surface irregularity ($R_a$) are communicated in

*Table 1.6* Relationship between experimental and predicted values of surface roughness

| *Trial no.* | *Prediction value* | *Experimental value* | *% e (error)* |
|---|---|---|---|
| 1 | 1.746 | 1.577 | 9.679 |
| 2 | 1.959 | 1.830 | 6.585 |
| 3 | 3.908 | 4.000 | −2.354 |
| 4 | 2.173 | 2.140 | 1.519 |
| 5 | 1.941 | 1.810 | 6.749 |
| 6 | 2.924 | 2.820 | 3.557 |
| 7 | 1.941 | 1.893 | 2.473 |
| 8 | 2.094 | 1.890 | 9.742 |
| 9 | 2.656 | 2.540 | 4.367 |
| 10 | 2.606 | 2.656 | −1.919 |
| 11 | 1.995 | 2.100 | −5.263 |
| 12 | 2.070 | 2.090 | −0.966 |
| 13 | 1.659 | 1.736 | −4.641 |
| 14 | 2.636 | 2.656 | −0.759 |
| 15 | 2.723 | 2.620 | 3.783 |
| 16 | 2.506 | 2.303 | 8.100 |
| 17 | 1.513 | 1.570 | −3.767 |
| 18 | 2.792 | 2.673 | 4.262 |

Table 1.6. On the support of outcome, it can be seen that the highest and lowest fault percentage for $R_a$ is 9.742% and −5.263%, which is suitable to a great extent. It has been found that approximate surface irregularity ($R_a$) is well within the limits. Hence, this model is used for optimization using GSA, PSO, and hybrid PSOGSA.

## 1.6 OPTIMIZATION

For optimization, a program is written in MATLAB. The achievement of the optimization method depends upon the choice of the parameters. A number of experiments are done to tune these parameters. Thirty runs are performed and the best result among the 30 is chosen as an ending solution. The parameter set for these three algorithms are given.

### 1.6.1 Setting of Parameters for GSA

Parameters used for application of GSA are

Population size = 100

Maximum iterations (T) = 500

$\alpha = 20$

Initial value of G0 = 100 that decreases to 1 according to Equation (1.8).

*Table 1.7* Optimum parameters to minimize surface roughness

| *Technique* | *Optimum parameters* | | | *Surface roughness* |
|---|---|---|---|---|
| | *Feed rate* | *Cutting speed* | *DOC* | |
| **GSA** | 0.0877 | 77.0789 | 0.2001 | 1.3750 |
| **PSO** | 0.0866 | 79.2958 | 0.2000 | 1.3744 |
| **PSOGSA** | 0.0866 | 78.2958 | 0.2000 | 1.3744 |

### 1.6.2 Setting of Parameters for PSO

Parameters used for PSO are:
Population size = 30
Maximum iterations (T) = 200
$c_1 = 2$
$c_2 = 2$
$w_{max} = 0.9$
$w_{min} = 0.4$
$w = w_{max} - (w_{max} - w_{min}) * t/T$
Where T: highest iterations
t: the present iteration

### 1.6.3 Setting of Parameters for PSOGSA

Parameters used for PSOGSA are:
Population size N = 30
Max_Iteration = 100
c1 = 2
c2 = 2
$\alpha = 20$
The initial value of G0 = 100 decreases to 1 with time. The optimum parameter for minimum surface roughness is shown in Table 1.7. Figure 1.6(a–c) shows the surface roughness ($R_a$) versus number of iterations for GSA, PSO, and PSOGSA. All three techniques gave about the same results. But the number of iterations required by PSOGSA is less, followed by PSO and GSA.

## 1.7 CONFIRMATION OF RESULTS

The reason of the verification test in this research is to confirm the most favorable cutting situation (x1 at level 2, x2 at level 3, x3 at level 2, x4 at level 2, x5 at level 2, and x6 at level 1) that is recommended with the trial. At best possible values, significant parameters are set at an optimum and the insignificant factors are set at an economic level. The confirmation results are compared by the predictable average. The predictable mean at the

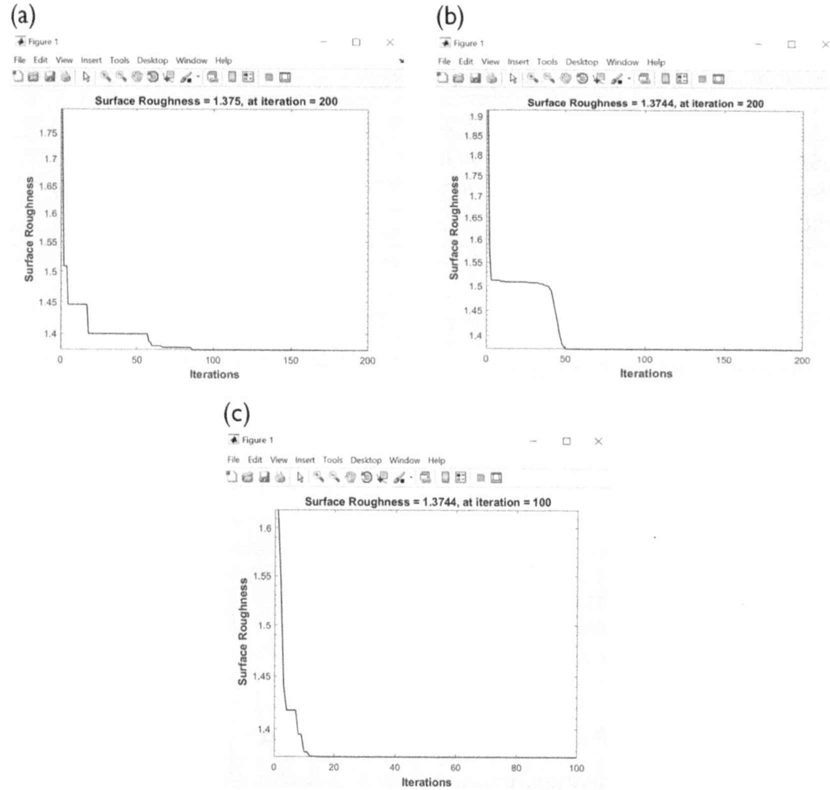


*Figure 1.6* Surface roughness versus iterations plot (a) GSA, (b) PSO, (c) PSOGSA.

most favorable setting could be obtained from Equation (1.23). A confidence interval on confirmation run for the predicted mean is calculated using Equation (1.24) (Ross, 1988):

$$\mu_{Ra} = x3(L2) + x4(L2) + x6(L1) - 2 * T_{Ra} \tag{1.23}$$

$$CI_{CE} = \sqrt{F_a(1, f_e)\, V_e \left[\frac{1}{n_{eff}} + \frac{1}{R}\right]} \tag{1.24}$$

wherever $F_\alpha$; $(1, f_e) = F_{0.05}$; $(1; 42) = 4.08$ (obtained from table), α (threat) = 0·05,

DOF associated with the $\mu_{Ra}$ = 11, Number of trials = 18, N = 18 × 3 = 54

neff = efficient replication = N/ {1 + [Total DOF related in the approximation of mean]} = 54 / (1 + 11) = 4.5

R = numeral of repetition for the trial = 3

Where $T_{Ra}$ = 2.272 (Table 1.3).

*Table 1.8* Parameters and their chosen levels

| *Process parameters* | *Optimal levels* |
|---|---|
| Tool nose radius (x1) | 0.8 mm ( the irrelevant variable put at profitable levels) |
| Rake angle (x2) | 0 degree (put at profitable levels) |
| Feed (x3) | 0.1 mm/rev |
| Speed (x4) | 110.84 m/min |
| Cutting environment (x5) | Wet (set at profitable levels) |
| DOC (x6) | 0.8 mm |

*Table 1.9* Confirmatory experimental results

| *Performance characteristics* | *Best levels of process parameters* | *Predicted value* | *Optimal experimental value (average of three verification experiment)* | *Predicted CI at 95 % CL ($CI_{CE}$)* | *Experimental value* |
|---|---|---|---|---|---|
| Surface Roughness | x1(L2),x2(L3), x3(L2), x4,(L2), x5(L2),x6(L1) | 1.385 μm | 1.650 μm | 0.761 < $\mu_{Ra}$ (microns) < 2.009 | 1.570 μm |

x3 (L2), x4 (L2). and x6 (L1) are the mean values of surface irregularity with parameters at the best possible levels. From response Table 1.4, $\overline{x3\ (L2),}$ = 1.976, $\overline{x4\ (L2)}$ = 2.006, and $\overline{x6\ (L1)}$ = 1.947. Hence, $\overline{\mu_{Ra}}$ = 1.385 μm, $\overline{Ve}$ = 0.1717 (Table 1.5).

A confidence internal (CI) for the predicted mean on a verification run is ± 0.624 using Equation (1.24) i.e. $CI_{CE}$ = ± 0.624.

The 95% CI of the predicted best surface roughness is: $[\mu_{Ra} - CI] < \mu_{Ra} < [\mu_{Ra} + CI]$ i.e. is $0.761 < \mu_{Ra}$ (microns) $< 2.009$.

Table 1.8 shows the parameters at the selected levels for optimization of surface roughness. The optimum average value of $R_a$ = 1.650 μm was obtained under the earlier mentioned cutting conditions on the lathe machine. Table 1.9 shows the confirmatory experimental results. The mean of the responses is established to be within the confidence interval. Table 1.3 shows the results are very closely related to the minimum average surface roughness ($R_a$) of 1.570 μm and individual quality characteristics of 1.38 μm in trial number 17. Table 1.7 shows that the three techniques (GSA, PSO, and PSOGSA) gave near about same results.

## 1.8 CONCLUSIONS

In this research, the PSO, GSA, and PSOGSA are used to optimize the process parameters in the machining of unidirectional-GFRP composite.

Surface roughness ($R_a$) is observed as an output quality characteristic. The experiments are conducted as per ($L_{18}$) orthogonal array. The following conclusions are made:

1. The ANOVA of surface roughness ($R_a$) model shows that the predictive model is accurate and credible. The ANOVA outcome shows that x6 (DOC), x4 (speed), and x3 (feed) have important effects on surface irregularity. x5, x2, and x1 have no consequence at 95% CL. The feed rate is found to be a very important factor while DOC is the least important factor. The feed rate is responsible for the roughness produced on the unidirectional-GFRP workpiece.
2. The percent contributions of DOC (10.584%), speed (21.595%), and feed (29.110%) are large, as compared to the other factors.
3. Multiple regression coefficients $R^2$ value is found be: 95.8%. $R^2$ value for the developed model using regression modeling is extremely sufficient as their $R^2$ value is very close to 1.
4. The optimization method used in this work can be used to find the necessary conditions to achieve the best quality under the given constraint. In this study, GSA, PSO, and PSOGSA are utilized to optimize machining parameters for a single objective in the machining of UD-GFRP. The experimental results show that PSOGSA has the advantage in convergence speed.
5. The minimum surface roughness obtained by PSOGSA is 1.374 μm at a feed (0.0866 mm/rev), DOC (0.20 mm), and speed (78.2958 m/min).
6. The analysis of results shows that minimum surface roughness can be attained by using a parametric value obtained by the PSOGSA algorithm.

## REFERENCES


Abrate S. & Walton D. A. (1992). Machining of composite materials (a two part review). *Composites Mfg*, 3(2), 75–94.

Battle, M. O. (2008). *Introduction to Production engineering Rio de Janeiro*. Elsevier Publisher.

Chatterjee, G., Mahanti, K. & Mahapatra, P. R. S. (2011). Generation of phase-only pencil-beam pair from concentric ring array antenna using gravitational search algorithm. *International Conference on Communications and Signal Processing*, 384–388.

Davim, J. P. & Reis, P. (2004). Multiple regression analysis (MRA) in modeling milling of glass fibre reinforced plastics (GFRP). *International Journal Manufacturing Technology Manag*, 6(1–2), 85–197.

Esmin, A. A., Lambert-Torres, G. & de Souza, A. C. Z. (2005). A hybrid particle swarm optimization applied to loss power minimization. *IEEE Transactions on Power Systems*, 20(2), 859–866.

Gao, Y., Li, Y. & Qian, H. (2008). The design of IIR digital filter based on chaos particle swarm optimization algorithm *IEEE International Conference on Genetic and Evolutionary Computing*, 303–306.

Goswami, D. & Chakraborty, S. (2015). Parametric optimization of ultrasonic machining process using gravitational search and fireworks algorithms. *Ain Shams Engineering Journal*, 6, 315–331.

Hussain, S. A., Pandurangadu, V. & Palanikumar, K. (2011). Machinability of glass fiber reinforced plastic (GFRP) composite materials. *International Journal of Engineering, Science and Technology*, 3(4), 103–118.

Isık, B. (2008). Experimental investigations of surface roughness in orthogonal turning of unidirectional glass-fiber reinforced plastic composite. *International Journal Advance Manufacturing Technology*, 37, 42–48.

Kolahan, F. & Abachizadeh, M. (2008). Optimizing turning parameters for cylindrical parts using simulated annealing method. *World Academy of Science, Engineering and Technology*, 46, 436–439.

Konig, W. C., Wulf, P. & Grab, W. H. (1985). Machining of fibre reinforced plastics. *CIRP Annals*, 34, 537–548.

Kumar, S., Meenu & Satsangi, P. S. (2013). Multiple-response optimization of turning machining by the taguchi method and the utility concept using unidirectional glass fiber-reinforced plastic composite and carbide (k10) cutting tool. *Journal of Mechanical Science and Technology*, 27 (9), 2829–2837.

Mirjalili, S., Mirjalili, S. M. & Lewis, A. (2014). Grey wolf optimizer. *Advances in Engineering Software*, 69, 46–61.

Mirjalili, S., Zaiton, S. & Hashim, M. (2010). A New Hybrid PSOGSA Algorithm for Function Optimization, *International Conference on Computer and Information Application*, 374–377.

Montogomery, D. C., Peck, E. A. & Vining, G. G. (2001). Introduction to linear regression analysis. John Wiley & Sons.

Palanikumar, L., Karunamoorthy, R., Karthikeyan, R., & Latha, B. (2006). Optimization of machining parameters in turning GFRP composites using a carbide (K10) tool based on the Taguchi method with fuzzy logics. *Metals and materials International*, 12(6), 483–491.

Raj, P., Elaya Perumal, A. & Ramu, P. (2012). Prediction of surface roughness and delamination in end milling of GFRP using mathematical model and ANN. *Indian Journal of Engineering & Materials Sciences*, 19, 107–120.

Rashedi, E., Nezamabadi-Pour, H. & Saryazdi, S. (2009). GSA: A gravitational search algorithm. *Information Sciences*, 179(13), 2232–2248.

Rashedi, E., Nezamabadi-Pour, H. & Saryazdi, S. (2011). Filter modeling using gravitational search algorithm. *Engineering Applications of Artificial Intelligence*, 24(1), 117–122.

Rather, S. A. & Shanthi Bala, P. (2020). Swarm-based chaotic gravitational search algorithm for solving mechanical engineering design problems. *World Journal of Engineering, Emerald Publishing Limited.*

Ross, P. J. (1988). *Taguchi techniques for quality engineering*. New York: McGraw-Hills Book Company.

Ross, P. J. (1996). *Taguchi techniques for quality engineering*. New York: McGraw-Hill Book Company.

Santhanakrishnan, G., Krishnamurthy, R. & Malhotra, S. K. (1989). High speed steel tool wear studies in machining of glass fibre-reinforced plastics. *Wear*, 132, 327–336.

Schutz, B. (2003). *Gravity from the ground up*. Cambridge, United Kingdom: Cambridge University Press.

Suman, B. & Kumar, P. A. (2006). Survey of simulated annealing as tool for single and multi-objective optimization. *Journal of the Operational Research Society*, 57(10), 1143–1160.

Yang, S., Srinivas, J., Mohan, S., Lee, D. & Balaji, S. (2009). Optimization of electric discharge machining using simulated annealing. *Journal of Materials Processing Technology*, 209, 4471–4475.

Chapter 2

# Multi-Response Optimization During High-Speed Drilling of Composite Laminate Using Grey Entropy Fuzzy (GEF) and Entropy-Based Weight Integrated Multi-Variate Loss Function

*Jalumedi Babu*[1], *Khaleel Ahmed*[1], *Lijo Paul*[2], *Abyson Scaria*[2], *and J. Paulo Davim*[3]

[1]Department of Mechanical Engineering, IMPACT college of Engineering & Applied Sciences, Bangalore, Karnataka, India

[2]Department of Mechanical Engineering, St. Joseph's College of Engineering & Technology, Kottayam, Kerala, India

[3]Department of Mechanical Engineering, University of Aveiro Campus, Santiago, Aveiro, Portugal

**CONTENTS**



DOI: 10.1201/9781003258421-3

## 2.1 INTRODUCTION

There is ongoing research on composite materials, utilising the advantages of high-speed machining. However, the quantity of research with high-speed drilling is less compared with the research on conventional low-speed drilling. The next paragraphs present a brief review on research on drilling composite materials with conventional low-speed drilling, followed by high-speed drilling.

Compared to conventional engineering materials utilised in the automobile, defence, and aviation manufacturing sectors, designers prefer composite materials because they are lightweight, stiff, and robust. [1]. Drilling and other machining operations are used for structural connecting. Drilling activities, particularly high-speed drilling, are problematic with these materials due to their great heterogeneity and low heat conductivity. Defects such as fibre pull out, delamination, matrix cracking, and debonding are prevalent [2–5]. The majority of studies in the literature concentrated on thrust force as well as its impact on drilling flaws, particularly delamination. Complete details on delamination, as well as methods for measuring and assessing it, can be found elsewhere [6–8]. Khashaba et al., for example, looked at the effect of machining settings during GFRPC drilling both theoretically and empirically, focusing on thrust force and delamination [9,10]. The effect of production procedures on the mechanical characteristics of GFRPC composites was investigated by Formisano et al. [11]. Erturk et al. [12] investigated the effect of drilling circumstances such as feed rate, spindle speed, and drill bit shape on GFRP composite temperature and delamination. Non-traditional strategies for reducing delamination during machining for GFRP composites were also discussed in the literature [13]. All of these tests were carried out using traditional low-speed drilling techniques.

There has been little research of high-speed drilling fibre-reinforced composite laminates, which allows for a faster material removal rate, according to the authors. Delamination is reduced and drilling costs are reduced with high-speed drilling. Researchers looked into the impact of high-speed drilling of delamination in a variety of methods. Kao and colleagues [14] studied micro-drill coatings and discovered that they increase hole quality. Lin et al. [15] investigated the effect of thrust force and torque on hole quality and determined that when spindle speed increases, cutting force drops, and delamination decreases. Rubio et al. [16] found that high-speed drilling resulted in better hole quality. Sanjay et al. [17] provided machinability maps of high-speed composites drilling, showing the effect of feed and spindle rotating speed on delamination, circularity error, hole size, and surface roughness. They concluded that such a machinability map may be utilised to determine the impact of machining force fluctuations on tool wear. The drilling settings were optimised by Vijayan Krishnaraj et al. [18], who found that feed had a significant effect on thrust and, thus, on delamination damage. The effect of feeds and spindle rotational speeds on delamination during high-speed machining has previously been observed elsewhere [19,20]. Although recent research on high-speed machining in a metal matrix [21] as well as CFRP composites [22] has concentrated on drilling delamination, the major goal of this study was to see how spindle rotating speed, as well as feed, affected delamination, torque, thrust force, and hole diameter when drilling GFRPC laminates.

This chapter presents an innovative methodology of combining grey relational analysis (GRA) with fuzzy logics, and also presently proposed entropy-based weight integrated multi-variate loss function for optimisation of drilling variables during high-speed machining of composite materials. Results obtained from both the methods are compared for the applicability of these innovative approaches.

## 2.2 MATERIALS AND METHODS

As indicated in Figure 2.1(a), drilling experiments were performed on a Vertical Machining Unit MAKINO S33 with a power of 11 KW and a spindle speed of up to 20,000 rpm. To make drilling testing easier, a composite laminate was put on a dynamometer to measure torque and thrust force, which was then arrested to the machine's bed, as shown in Figure 2.1(b). Drill bits utilised were two-flute coated carbide (K20) drill bits with a diameter of 5 mm. This study's GFRP sheet was comprised of woven fibres reinforced in an epoxy matrix. The composite was 250 × 35 × 3 $mm^3$ in size and 1.9 $g/cm^3$ in density. Two machining conditions were used in the experiment: feed and spindle rotation. There are five levels to each control factor. To ensure dependability, each experiment is done twice. The response is the average of these tests. The feeds were 500, 1,000, 3,000, 6,000, and 9,000 mm/min, while the spindle rotational speeds were 10,000–18,000 rpm (in the steps of 2,000). Experiments were carried out using the orthogonal array L-25, as shown in Table 2.1. A dynamometer

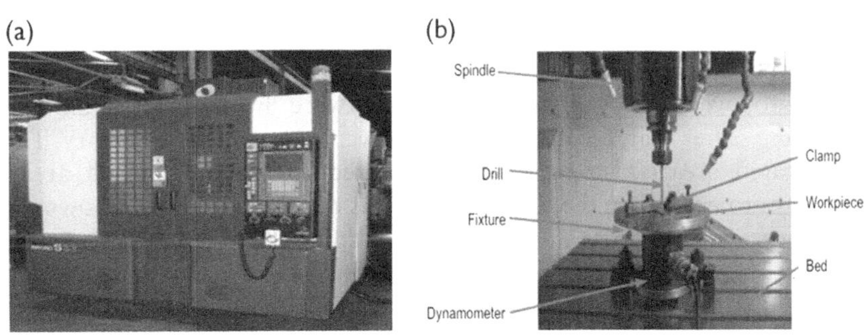


*Figure 2.1* (a) Vertical machining centre, (b) dynamometer arrangement for measuring thrust force.

*Table 2.1* Machining conditions

| *Drilling of GFRP laminate* | |
|---|---|
| Equipment | Makino S33 VMC, with power 11 kW |
| Workpiece | GFRP 3 mm thick |
| Drilling tool | Two-flute, 5 mm diameter coated carbide (K20) drill. |
| Drilling conditions | Spindle speed in rpm: 10,000, 12,000, 14,000, 16,000, and 18,000; feed rate (mm/min): 500, 1,000, 3,000, 6,000, and 9,000. |

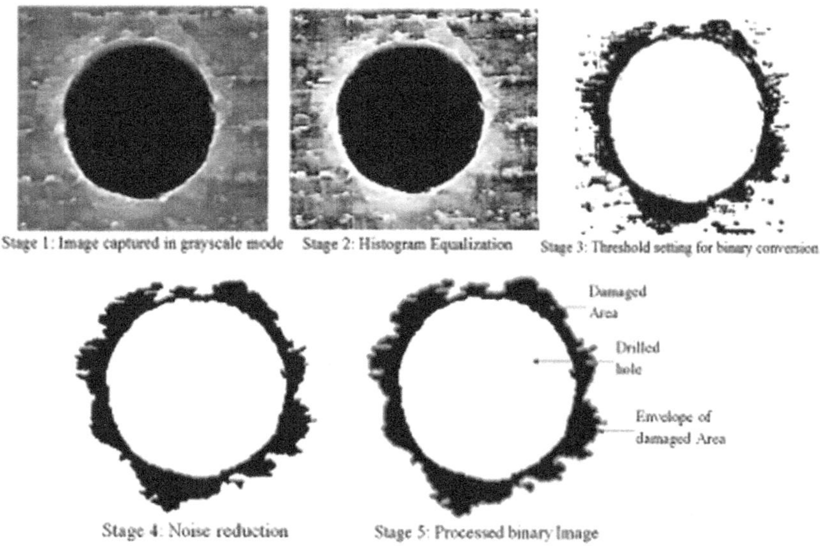


*Figure 2.2* Image processing procedure.

(Syscon SI-223D) was used to measure thrust forces during milling. During drilling, a digital storage recorder was utilised to measure and store the forces. Delamination was assessed via scanning the holed sheet using a flat-bed scanner at a resolution of 1,200 PPI, and then saving the bitmap image. The photos were then transferred to the software 'Image J' for additional processing. After that, colour photographs were transformed to binary. The delaminated area's histogram was compared to the undamaged area's histogram to determine the threshold for this change. Figure 2.2 shows the image processing approach used to obtain a delaminated area of acceptable quality. Delamination was determined using an adjusted delamination factor. Equations for the delaminations were previously published [6]. The hole diameter was measured using a coordinate measuring machine. The output values were analysed and optimised using Minitab 20 and MATLAB R2021a.

## 2.3 RESULTS AND DISCUSSIONS

Experimental responses i.e., thrust force, torque, delamination both push-out and peel-up, and hole diameter at different machining conditions are shown in Table 2.2.

Table 2.2 shows that as spindle rotation increases, delamination values increase as well. Krishnaraj et al. [18] found a similar pattern in their findings. The reason being that the increased heat generated during drilling with extremely high spindle revolutions soothes the composite matrix. This weakens the laminate and causes to the emergence of delamination at significantly lower cutting forces. It's also worth noting that the value of exit delamination is higher for smaller feeds, such as 500 mm/min. Because of the limited heat transfer of the soft epoxy matrix, any heat created by high

*Table 2.2* Results

| *Experiment no.* | *Spindle speed (rpm)* | *Feed rate (mm/min)* | *Thrust force (N)* | *Torque (Nm)* | *Delamination factor* | | *Hole diameter (mm)* |
|---|---|---|---|---|---|---|---|
| | | | | | *Peel-up* | *Push-out* | |
| 1 | 10000 | 500 | 31 | 0.05 | 1.633 | 1.499 | 5.01 |
| 2 | 10,000 | 1,000 | 35 | 0.07 | 1.112 | 1.113 | 5.02 |
| 3 | 10,000 | 3,000 | 42 | 0.1 | 1.55 | 1.399 | 5.01 |
| 4 | 10,000 | 6,000 | 51 | 0.2 | 1.949 | 1.687 | 5.00 |
| 5 | 10,000 | 9,000 | 60 | 0.28 | 2.184 | 2.191 | 5.01 |
| 6 | 12,000 | 500 | 26 | 0.05 | 2.272 | 1.77 | 5.42 |
| 7 | 12,000 | 1,000 | 32 | 0.06 | 1.207 | 1.061 | 5.03 |
| 8 | 12,000 | 3,000 | 36 | 0.1 | 1.374 | 1.343 | 5.02 |
| 9 | 12,000 | 6,000 | 42 | 0.18 | 1.818 | 1.581 | 5.00 |
| 10 | 12,000 | 9,000 | 54 | 0.22 | 2.154 | 2.085 | 5.00 |
| 11 | 14,000 | 500 | 20 | 0.04 | 2.545 | 2.443 | 5.58 |
| 12 | 14,000 | 1,000 | 23 | 0.06 | 1.801 | 1.71 | 5.36 |
| 13 | 14,000 | 3,000 | 31 | 0.09 | 1.7 | 1.596 | 5.08 |
| 14 | 14,000 | 6,000 | 37 | 0.14 | 1.93 | 1.855 | 5.01 |
| 15 | 14,000 | 9,000 | 45 | 0.18 | 2.234 | 2.307 | 5.05 |
| 16 | 16,000 | 500 | 17 | 0.03 | 2.647 | 3.179 | 5.64 |
| 17 | 16,000 | 1,000 | 20 | 0.06 | 2.163 | 2.186 | 5.85 |
| 18 | 16,000 | 3,000 | 25 | 0.09 | 1.897 | 1.944 | 5.05 |
| 19 | 16,000 | 6,000 | 33 | 0.1 | 1.992 | 1.94 | 5.03 |
| 20 | 16,000 | 9,000 | 39 | 0.15 | 2.186 | 2.419 | 5.04 |
| 21 | 18,000 | 500 | 16 | 0.03 | 2.954 | 3.391 | 5.74 |
| 22 | 18,000 | 1,000 | 20 | 0.05 | 2.619 | 2.496 | 5.90 |
| 23 | 18,000 | 3,000 | 23 | 0.08 | 2.4 | 2.408 | 5.44 |
| 24 | 18,000 | 6,000 | 32 | 0.08 | 1.931 | 2.108 | 5.21 |
| 25 | 18,000 | 9,000 | 36 | 0.11 | 2.122 | 2.51 | 5.23 |

spindle speeds is concentrated near the hole [20]. The machinability maps of Rawat and Attia [17] also confirm the same. Rawat and Attia's machinability maps [17] also support this. These maps show that delamination is strong at very low feed rates, such as 20 mm/rev, with a high spindle rotation of 15,000 rpm; however, as feed rates increased to higher levels, such as 60 and 100 mm/rev, delamination values reduced.

Table 2.2 also demonstrates that as the spindle speed increases, the thrust force values drop. This could be due to increased heat generation during drilling with greater spindle revolutions, which then soothes the matrix, as previously mentioned. Thrust force and torque, on the other hand, rise as the feed increases, as larger feeds necessitate greater forces as well as torque for drilling.

Due to the obvious thermo-mechanical characteristics of the GFRP laminates at higher spindle speeds at lower feed values, hole sizes increased with the increase in spindle rotations and reduced with the rise in feeds, as shown in Table 2.2 [20].

To obtain optimum machining conditions for these multi-responses simultaneously two innovative approaches, combining grey relational analysis (GRA) with fuzzy logics and also presently proposed entropy-based weight integrated multi-Vvriate loss function are used for optimisation of drilling variables during high-speed machining of composite materials.

## 2.4 OPTIMIZATION WITH ENTROPY WEIGHT-BASED GREY RELATIONAL ANALYSIS

### 2.4.1 Grey Relational Analysis

This problem of high-speed drilling during machining of GFRP laminate is optimised with GRA, grey relational analysis. GRA is one quantitative method for obtaining the inequality and equality between the input conditions. Because each of the output responses has different ranges and units, it is required to normalize the values of the responses in the range between 1 and 0. In multi-objective optimisation problems, it may be necessary to consider some responses as maximisation (higher the better), some responses as minimisation (smaller the better), and some may be as nominal value is better type. GRA converts multi-response problems in to a single response called the grey relational coefficient.

Responses of maximisation are normalised by using Equation (2.1).

$$X_i^*(k) = \frac{X_i^0(k) - minX_i^0(k)}{maxX_i^0(k) - \min X_i^0(k)} \tag{2.1}$$

Responses of minimisation are normalised by using Equation (2.2).

$$X_i^*(k) = \frac{max\ X_i^0(k) - X_i^0(k)}{maxX_i^0(k) - \min X_i^0(k)} \tag{2.2}$$

Responses of nominal the better are normalised by using Equation (2.3).

$$X_i^* = 1 - \frac{|X_i^0(k) - X^0|}{\max X_i^0(k) - X_i^0} \tag{2.3}$$

i = 1, 2, 3 … , n,
n is the number of machining parameters
m is the number of experimental data
$X_i^0(k)$ is the original sequence
$X_i^*(k)$ is the sequence after data pre-processing
max $X_i^0(k)$ is the largest value in $X_i^0(k)$
min $X_i^0(k)$ is the smallest value in $X_i^0(k)$
$X^0$ is the desired value
Normalised values calculated for all the responses are tabulated in Table 2.3.

## 2.4.2 Grey Relational Coefficient

Now, the grey relational coefficient is determined by using Equation (2.4).

$$\xi(k) = \frac{\Delta_{min} + \zeta\Delta_{max}}{\Delta_{0i}(k) + \zeta\Delta_{max}} \tag{2.4}$$

$\Delta_{0i}(k)$, $\Delta_{max}$, $\Delta_{min}$ are calculated using Equations (2.5), (2.6), and (2.7).

$$\Delta_{0i}(k) = \|X_0^*(k) - X_i^*(k)\| \tag{2.5}$$

$$\Delta_{max} = \max\max \|X_0^*(k) - X_i^*(k)\| \tag{2.6}$$

$$\Delta_{min} = \min\min \|X_0^*(k) - X_i^*(k)\| \tag{2.7}$$

$\zeta$ is the identification coefficient, and its ranges are $\zeta$ to [0, 1]. generally, $\zeta = 0.5$ is used.

Where $\Delta_{0i}(k)$ is the deviation sequence for the reference sequence $X_0^*(k)$
$X_i^*(k)$ is the comparability sequence

The deviation sequence can be obtained after data preprocessing, shown in Table 2.3. The values $\Delta_{max}$ and $\Delta_{min}$ can be found from Table 2.3. Grey relational coefficients obtained with Equation (2.4) for all drilling conditions are shown in Table 2.3.

*Table 2.3* Normalised and deviation sequence for the responses

| *Experiment no.* | *Normalised values* | | | | | *Deviation sequence* | | | | |
|---|---|---|---|---|---|---|---|---|---|---|
| | *A* | *B* | *C* | *D* | *E* | *A* | *B* | *C* | *D* | *E* |
| 1 | 0.6591 | 0.9200 | 0.6760 | 0.7913 | 0.9889 | 0.3409 | 0.0800 | 0.3240 | 0.2087 | 0.0111 |
| 2 | 0.5682 | 0.8400 | 0.9427 | 0.9527 | 0.9778 | 0.4318 | 0.1600 | 0.0573 | 0.0473 | 0.0222 |
| 3 | 0.4091 | 0.7200 | 0.7185 | 0.8331 | 0.9889 | 0.5909 | 0.2800 | 0.2815 | 0.1669 | 0.0111 |
| 4 | 0.2045 | 0.3200 | 0.5143 | 0.7127 | 0.9889 | 0.7955 | 0.6800 | 0.4857 | 0.2873 | 0.0111 |
| 5 | 0.0000 | 0.0000 | 0.3941 | 0.5019 | 0.9889 | 1.0000 | 1.0000 | 0.6059 | 0.4981 | 0.0111 |
| 6 | 0.7727 | 0.9200 | 0.3490 | 0.6780 | 0.5333 | 0.2273 | 0.0800 | 0.6510 | 0.3220 | 0.4667 |
| 7 | 0.6364 | 0.8800 | 0.8941 | 0.9745 | 0.9667 | 0.3636 | 0.1200 | 0.1059 | 0.0255 | 0.0333 |
| 8 | 0.5455 | 0.7200 | 0.8086 | 0.8565 | 0.9778 | 0.4545 | 0.2800 | 0.1914 | 0.1435 | 0.0222 |
| 9 | 0.4091 | 0.4000 | 0.5814 | 0.7570 | 1.0000 | 0.5909 | 0.6000 | 0.4186 | 0.2430 | 0.0000 |
| 10 | 0.1364 | 0.2400 | 0.4094 | 0.5462 | 1.0000 | 0.8636 | 0.7600 | 0.5906 | 0.4538 | 0.0000 |
| 11 | 0.9091 | 0.9600 | 0.2093 | 0.3965 | 0.3556 | 0.0909 | 0.0400 | 0.7907 | 0.6035 | 0.6444 |
| 12 | 0.8409 | 0.8800 | 0.5901 | 0.7031 | 0.6000 | 0.1591 | 0.1200 | 0.4099 | 0.2969 | 0.4000 |
| 13 | 0.6591 | 0.7600 | 0.6418 | 0.7507 | 0.9111 | 0.3409 | 0.2400 | 0.3582 | 0.2493 | 0.0889 |
| 14 | 0.5227 | 0.5600 | 0.5241 | 0.6424 | 0.9889 | 0.4773 | 0.4400 | 0.4759 | 0.3576 | 0.0111 |
| 15 | 0.3409 | 0.4000 | 0.3685 | 0.4534 | 0.9444 | 0.6591 | 0.6000 | 0.6315 | 0.5466 | 0.0556 |
| 16 | 0.9773 | 1.0000 | 0.1571 | 0.0887 | 0.0556 | 0.0227 | 0.0000 | 0.8429 | 0.9113 | 0.9444 |
| 17 | 0.9091 | 0.8800 | 0.4048 | 0.5040 | 0.2889 | 0.0909 | 0.1200 | 0.5952 | 0.4960 | 0.7111 |
| 18 | 0.7955 | 0.7600 | 0.5409 | 0.6052 | 0.9444 | 0.2045 | 0.2400 | 0.4591 | 0.3948 | 0.0556 |
| 19 | 0.6136 | 0.7200 | 0.4923 | 0.6069 | 0.9667 | 0.3864 | 0.2800 | 0.5077 | 0.3931 | 0.0333 |
| 20 | 0.4773 | 0.5200 | 0.3930 | 0.4065 | 0.9556 | 0.5227 | 0.4800 | 0.6070 | 0.5935 | 0.0444 |
| 21 | 1.0000 | 1.0000 | 0.0000 | 0.0000 | 0.0000 | 0.0000 | 0.0000 | 1.0000 | 1.0000 | 1.0000 |
| 22 | 0.9091 | 0.9200 | 0.1714 | 0.3743 | 0.1778 | 0.0909 | 0.0800 | 0.8286 | 0.6257 | 0.8222 |
| 23 | 0.8409 | 0.8000 | 0.2835 | 0.4111 | 0.5111 | 0.1591 | 0.2000 | 0.7165 | 0.5889 | 0.4889 |
| 24 | 0.6364 | 0.8000 | 0.5235 | 0.5366 | 0.7667 | 0.3636 | 0.2000 | 0.4765 | 0.4634 | 0.2333 |
| 25 | 0.5455 | 0.6800 | 0.4258 | 0.3685 | 0.7444 | 0.4545 | 0.3200 | 0.5742 | 0.6315 | 0.2556 |

The following notations are used for responses:

A. Thrust Force.
B. Torque.
C. Peel-up Delamination Factor.
D. Push-out Delamination Factor.
E. Hole Diameter.

### 2.4.3 Grey Relational Grade

Grey relational grade is generally calculated as a mean of all grey relational coefficients of corresponding responses, as in Equation (2.8). This equation is applicable only when equal weight is assigned to all the responses.

$$GRG = \frac{1}{n} \sum_{k=1}^{n} \zeta_i(\mathrm{k}) \tag{2.8}$$

For real-life problems like high-speed drilling, all the output responses are not equally important; hence, by giving different weights for the responses, Equation (2.8) can be altered as Equation (2.9).

$$GRG = \frac{1}{n} \sum_{k=1}^{n} W_k \, \zeta_i(k), \quad \sum_{k=1}^{n} W_k = 1 \tag{2.9}$$

Where $W_k$ shows the normalised weight of response k; for the equal weights, Equations (2.8) and (2.9) are the same.

In this work, an objective method, the entropy method [23–25], is applied for the calculation of the weights for the responses. The detailed procedure is presented below.

### 2.4.4 Entropy Method

This method was developed by Weaver and Shanon in 1947 and further refined by Zeleney in 1982 for calculating objective weights of individual responses. Entropy is basically an estimate of uncertainty in the data or information. It uses the theory of probability. Weights by the entropy method are determined by normalising data by using Equations (2.2) and (2.3) and the corresponding entropy values of each of the responses are calculated using Equation (2.10).

$$e_j = -\frac{1}{lnm} \sum_{i=1}^{m} P_{ij} \ln P_{ij} \tag{2.10}$$

Here, $P_{ij} = \frac{y_{ij}}{\sum_{i=1}^{m} y_{ij}}$ and m is the total number of experiments in this study, m = 25. The performance characteristic/response is decreasing or increasing

*Table 2.4* Calculation of weightages for the responses

| *Response* | $e_j$ | $(1 - e_j)$ | $w_j$ |
|---|---|---|---|
| A | 0.968 | 0.032 | 20 |
| B | 0.977 | 0.023 | 15 |
| C | 0.963 | 0.037 | 23 |
| D | 0.965 | 0.035 | 22 |
| E | 0.969 | 0.031 | 20 |
| Sum | | 0.158 | 100 |

with the entropy $(e_j)$. Now, individual weights can be calculated by Equation (2.11).

$$w_j = \frac{(1 - e_j)}{\sum_{i=1}^{n}(1 - e_j)} \tag{2.11}$$

Calculated normalized, *P, and* $P_{ij}$ in $P_{ij}$ values for the responses are shown in Table 2.4. Now, using these values, entropy and weights are calculated for the responses using Equations (2.10) and (2.11), respectively, and the values obtained are presented in Table 2.5.

Entropy value for the response A is $e_j = -\frac{1}{lnm}\sum_{i=1}^{m} P_{ij} = -\frac{1}{ln25}(-3.116) =$ 0.968.

The calculated weights using Equation (2.11) obtained for the responses are thrust force, torque, peel-up/entry delamination factor, push-out/exit delamination factor, and hole diameter are 0.20, 0.15, 0.23, 0.22, and 0.20, respectively. The same values are considered in the calculation of grey relational grade using Equation (2.9). Table 2.6 represents the obtained grey relational coefficients, grey relation grade, and the rank for all test runs.

### 2.4.5 Optimisation with GREG

Grey relational analysis with entropy-based weights, grey relational entropy grade, was determined. Experiment number 7 shows the maximum value of the GREG represents the optimum combination of input parameters. The GREG values varied from 0.5265 to 0.8234. The experimental values of all responses at optimised machining conditions are peel-up delamination factor 1.207, push-out delamination factor 1.061, thrust force 32 N, torque 0.06 Nm, and hole diameter 5.03 mm.

## 2.5 OPTIMISATION USING GREY ENTROPY FUZZY METHOD (GEFM)

The above-mentioned methods like entropy-based weight integrated multivariate loss function and also grey relation analysis with entropy-based

*Table 2.5* Normalised, $P_{ij}$, *and* $P_{ij}$ ln $P_{ij}$ values for the responses

| *Experiment no.* | *Normalised values* | | | | | $P_{ij}$ *values* | | | | | $P_{ij}$ *ln* $P_{ij}$ *values* | | | | |
|---|---|---|---|---|---|---|---|---|---|---|---|---|---|---|---|
| | *A* | *B* | *C* | *D* | *E* | *A* | *B* | *C* | *D* | *E* | *A* | *B* | *C* | *D* | *E* |
| 1 | 0.6667 | 0.9259 | 0.6835 | 0.7921 | 0.9900 | 0.0429 | 0.0510 | 0.0555 | 0.0547 | 0.0514 | −0.135 | −0.152 | −0.160 | −0.159 | −0.153 |
| 2 | 0.5778 | 0.8519 | 0.9440 | 0.9529 | 0.9800 | 0.0372 | 0.0469 | 0.0767 | 0.0658 | 0.0509 | −0.122 | −0.144 | −0.197 | −0.179 | −0.152 |
| 3 | 0.4222 | 0.7407 | 0.7250 | 0.8338 | 0.9900 | 0.0272 | 0.0408 | 0.0589 | 0.0575 | 0.0514 | −0.098 | −0.131 | −0.167 | −0.164 | −0.153 |
| 4 | 0.2222 | 0.3704 | 0.5255 | 0.7138 | 0.9900 | 0.0143 | 0.0204 | 0.0427 | 0.0493 | 0.0514 | −0.061 | −0.079 | −0.135 | −0.148 | −0.153 |
| 5 | 0.0222 | 0.0741 | 0.4080 | 0.5038 | 0.9900 | 0.0014 | 0.0041 | 0.0331 | 0.0348 | 0.0514 | −0.009 | −0.022 | −0.113 | −0.117 | −0.153 |
| 6 | 0.7778 | 0.9259 | 0.3640 | 0.6792 | 0.5800 | 0.0501 | 0.0510 | 0.0296 | 0.0469 | 0.0301 | −0.150 | −0.152 | −0.104 | −0.143 | −0.105 |
| 7 | 0.6444 | 0.8889 | 0.8965 | 0.9746 | 0.9700 | 0.0415 | 0.0490 | 0.0728 | 0.0673 | 0.0504 | −0.132 | −0.148 | −0.191 | −0.182 | −0.151 |
| 8 | 0.5556 | 0.7407 | 0.8130 | 0.8571 | 0.9800 | 0.0358 | 0.0408 | 0.0660 | 0.0591 | 0.0509 | −0.119 | −0.131 | −0.179 | −0.167 | −0.152 |
| 9 | 0.4222 | 0.4444 | 0.5910 | 0.7579 | 1.0000 | 0.0272 | 0.0245 | 0.0480 | 0.0523 | 0.0519 | −0.098 | −0.091 | −0.146 | −0.154 | −0.154 |
| 10 | 0.1556 | 0.2963 | 0.4230 | 0.5479 | 1.0000 | 0.0100 | 0.0163 | 0.0344 | 0.0378 | 0.0519 | −0.046 | −0.067 | −0.116 | −0.124 | −0.154 |
| 11 | 0.9111 | 0.9630 | 0.2275 | 0.3988 | 0.4200 | 0.0587 | 0.0531 | 0.0185 | 0.0275 | 0.0218 | −0.166 | −0.156 | −0.074 | −0.099 | −0.083 |
| 12 | 0.8444 | 0.8889 | 0.5995 | 0.7042 | 0.6400 | 0.0544 | 0.0490 | 0.0487 | 0.0486 | 0.0332 | −0.158 | −0.148 | −0.147 | −0.147 | −0.113 |
| 13 | 0.6667 | 0.7778 | 0.6500 | 0.7517 | 0.9200 | 0.0429 | 0.0429 | 0.0528 | 0.0519 | 0.0478 | −0.135 | −0.135 | −0.155 | −0.153 | −0.145 |
| 14 | 0.5333 | 0.5926 | 0.5350 | 0.6438 | 0.9900 | 0.0343 | 0.0327 | 0.0435 | 0.0444 | 0.0514 | −0.116 | −0.112 | −0.136 | −0.138 | −0.153 |
| 15 | 0.3556 | 0.4444 | 0.3830 | 0.4554 | 0.9500 | 0.0229 | 0.0245 | 0.0311 | 0.0314 | 0.0493 | −0.086 | −0.091 | −0.108 | −0.109 | −0.148 |
| 16 | 0.9778 | 1.0000 | 0.1765 | 0.0921 | 0.1500 | 0.0629 | 0.0551 | 0.0143 | 0.0064 | 0.0078 | −0.174 | −0.160 | −0.061 | −0.032 | −0.038 |
| 17 | 0.9111 | 0.8889 | 0.4185 | 0.5058 | 0.3600 | 0.0587 | 0.0490 | 0.0340 | 0.0349 | 0.0187 | −0.166 | −0.148 | −0.115 | −0.117 | −0.074 |
| 18 | 0.8000 | 0.7778 | 0.5515 | 0.6067 | 0.9500 | 0.0515 | 0.0429 | 0.0448 | 0.0419 | 0.0493 | −0.153 | −0.135 | −0.139 | −0.133 | −0.148 |
| 19 | 0.6222 | 0.7407 | 0.5040 | 0.6083 | 0.9700 | 0.0401 | 0.0408 | 0.0409 | 0.0420 | 0.0504 | −0.129 | −0.131 | −0.131 | −0.133 | −0.151 |
| 20 | 0.4889 | 0.5556 | 0.4070 | 0.4088 | 0.9600 | 0.0315 | 0.0306 | 0.0331 | 0.0282 | 0.0498 | −0.109 | −0.107 | −0.113 | −0.101 | −0.149 |
| 21 | 1.0000 | 1.0000 | 0.0230 | 0.0037 | 0.1000 | 0.0644 | 0.0551 | 0.0019 | 0.0003 | 0.0052 | −0.177 | −0.160 | −0.012 | −0.002 | −0.027 |
| 22 | 0.9111 | 0.9259 | 0.1905 | 0.3767 | 0.2600 | 0.0587 | 0.0510 | 0.0155 | 0.0260 | 0.0135 | −0.166 | −0.152 | −0.064 | −0.095 | −0.058 |
| 23 | 0.8444 | 0.8148 | 0.3000 | 0.4133 | 0.5600 | 0.0544 | 0.0449 | 0.0244 | 0.0285 | 0.0291 | −0.158 | −0.139 | −0.091 | −0.101 | −0.103 |
| 24 | 0.6444 | 0.8148 | 0.5345 | 0.5383 | 0.7900 | 0.0415 | 0.0449 | 0.0434 | 0.0371 | 0.0410 | −0.132 | −0.139 | −0.136 | −0.122 | −0.131 |
| 25 | 0.5556 | 0.7037 | 0.4390 | 0.3708 | 0.7700 | 0.0358 | 0.0510 | 0.0357 | 0.0256 | 0.0400 | −0.119 | −0.126 | −0.119 | −0.094 | −0.129 |
| | | | | Sum of $\sum_{i=1}^{m} P_{ij} \ln P_{ij}$ | | | | | | | −3.116 | −3.153 | −3.108 | −3.114 | −3.128 |

*Table 2.6* Grey relational coefficients with entropy-based weights (in brackets), grey relational grade, and rank for the various test runs

| *Experiment no.* | *Grey relational coefficients* | | | | | *Grey relational entropy grade* | *Rank* |
|---|---|---|---|---|---|---|---|
| | *Thrust force (0.20)* | *Torque (0.15)* | *Peel-up delamination (0.23)* | *Push-out delamination (0.22)* | *Hole diameter (0.20)* | | |
| 1 | 0.5946 | 0.8621 | 0.6068 | 0.7055 | 0.9783 | 0.7387 | 3 |
| 2 | 0.5366 | 0.7576 | 0.8972 | 0.9136 | 0.9574 | 0.8198 | 2 |
| 3 | 0.4583 | 0.6410 | 0.6398 | 0.7498 | 0.9783 | 0.6956 | 5 |
| 4 | 0.3860 | 0.4237 | 0.5073 | 0.6351 | 0.9783 | 0.5928 | 18 |
| 5 | 0.3333 | 0.3333 | 0.4521 | 0.5009 | 0.9783 | 0.5265 | 25 |
| 6 | 0.6875 | 0.8621 | 0.4344 | 0.6082 | 0.5172 | 0.6040 | 12 |
| 7 | 0.5789 | 0.8065 | 0.8252 | 0.9515 | 0.9375 | 0.8234 | 1 |
| 8 | 0.5238 | 0.6410 | 0.7232 | 0.7771 | 0.9574 | 0.7297 | 4 |
| 9 | 0.4583 | 0.4545 | 0.5443 | 0.6730 | 1.0000 | 0.6331 | 10 |
| 10 | 0.3667 | 0.3968 | 0.4585 | 0.5242 | 1.0000 | 0.5536 | 22 |
| 11 | 0.8462 | 0.9259 | 0.3874 | 0.4531 | 0.4369 | 0.5843 | 14 |
| 12 | 0.7586 | 0.8065 | 0.5495 | 0.6274 | 0.5556 | 0.6482 | 8 |
| 13 | 0.5946 | 0.6757 | 0.5826 | 0.6673 | 0.8491 | 0.6709 | 6 |
| 14 | 0.5116 | 0.5319 | 0.5123 | 0.5830 | 0.9783 | 0.6239 | 11 |
| 15 | 0.4314 | 0.4545 | 0.4419 | 0.4777 | 0.9000 | 0.5412 | 23 |
| 16 | 0.9565 | 1.0000 | 0.3723 | 0.3543 | 0.3462 | 0.5741 | 16 |
| 17 | 0.8462 | 0.8065 | 0.4565 | 0.5020 | 0.4128 | 0.5882 | 13 |
| 18 | 0.7097 | 0.6757 | 0.5213 | 0.5588 | 0.9000 | 0.6661 | 7 |
| 19 | 0.5641 | 0.6410 | 0.4962 | 0.5598 | 0.9375 | 0.6338 | 9 |
| 20 | 0.4889 | 0.5102 | 0.4517 | 0.4573 | 0.9184 | 0.5625 | 21 |
| 21 | 1.0000 | 1.0000 | 0.3333 | 0.3333 | 0.3333 | 0.5667 | 17 |
| 22 | 0.8462 | 0.8621 | 0.3763 | 0.4442 | 0.3782 | 0.5585 | 19 |
| 23 | 0.7586 | 0.7143 | 0.4110 | 0.4592 | 0.5056 | 0.5555 | 20 |
| 24 | 0.5789 | 0.7143 | 0.5121 | 0.5190 | 0.6818 | 0.5912 | 15 |
| 25 | 0.5238 | 0.6098 | 0.4655 | 0.4419 | 0.6618 | 0.5328 | 24 |

weight have limitations and uncertainty because of the use of nominal-the-better, higher-the-better, and lower-the-better characteristics for optimisation of processes like high-speed drilling. But, these methods are very efficient and effective methods for conversion of multi-objective problems into single-objective problems. Hence, further refinement of optimisation problems by genetic algorithm is used. Now the problem is converted to a single-objective function called grey relational entropy fuzzy with genetic algorithms.

Earlier optimisation studies in the literature give either the same weights for all the responses or use a subjected method like the SIMO method, which are inadequate/improper ways to optimise the process parameters. Hence, the entropy method, which is an objective method, is used along with GRA to calculate most scientific objective weights to transform a multi-objective problem to a single-objective named GREG. In this way, complex multi-objective optimisation problems can be solved with integration of grey relational analysis, entropy method, and fuzzy logic. Nonetheless, in this conversion process, there is disregard of relationships with evaluating responses. This method now presents the enhanced optimisation using correct weights for each response.

### 2.5.1 Grey Entropy Fuzzy Model

It was explained before that GREG includes a entropy-based weighted sum for all five grey relational coefficients; hence, for maximising, the GREG problem can be optimised without the consideration of responses criteria. For this, the fuzzy model was applied for this desirability function called GREG, considering all input factors. In a fuzzy model, the first step is fuzzification output and input data, the next step is generation of fuzzy rues, and then finally, de-fuzzification of results or outcomes.

The equivalent GEFG values are later derived using a fuzzy-logic toolbox of MATLAB (R2021a). The GREG values of torque, thrust force, peel-up, and push-out delamination factors and hole diameter are used as the inputs for fuzzy-logic system, the triangular membership functions used for the fuzzy modeling. Nine linguistic membership functions, as lowest (LT), very low (VL), medium low (ML), low (L), medium high (MH), high (H), medium higher (MHR), higher (HR), and highest (HT) are used for both input factors as well as output of the grey entropy fuzzy grade. These membership functions are presented in Figure 2.3. The obtained GEFG values are shown in Figure 2.4, as displayed in the fuzzy rule viewer. In this figure, 25 rows show the fuzzy rules used and the five columns represent GRC values of thrust force, torque, peel-up, and push-out delamination factors and hole diameter, and the last column gives the de-fuzzified GEFG values. These GEFG values captured for all 25 experiments are displayed in Table 2.7. It can be noted from this table that Experiment Number 7 (spindle speed of 12,000 rpm, feed rate of 1,000 mm/min) has the highest GEFG value, indicating the optimum combination of input factors for high drilling performance.

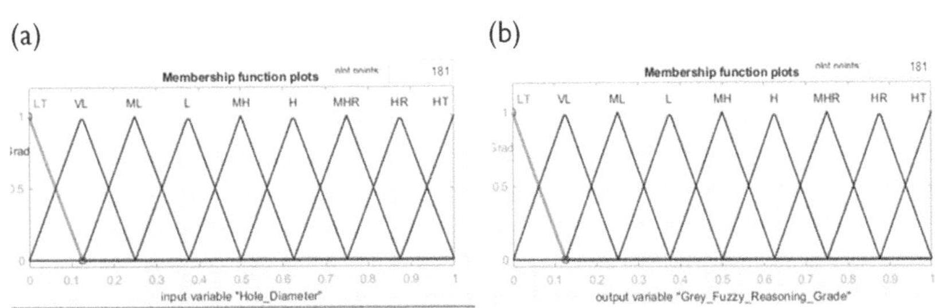


*Figure 2.3* Membership functions (a) for input factors, (b) for output GEFG.

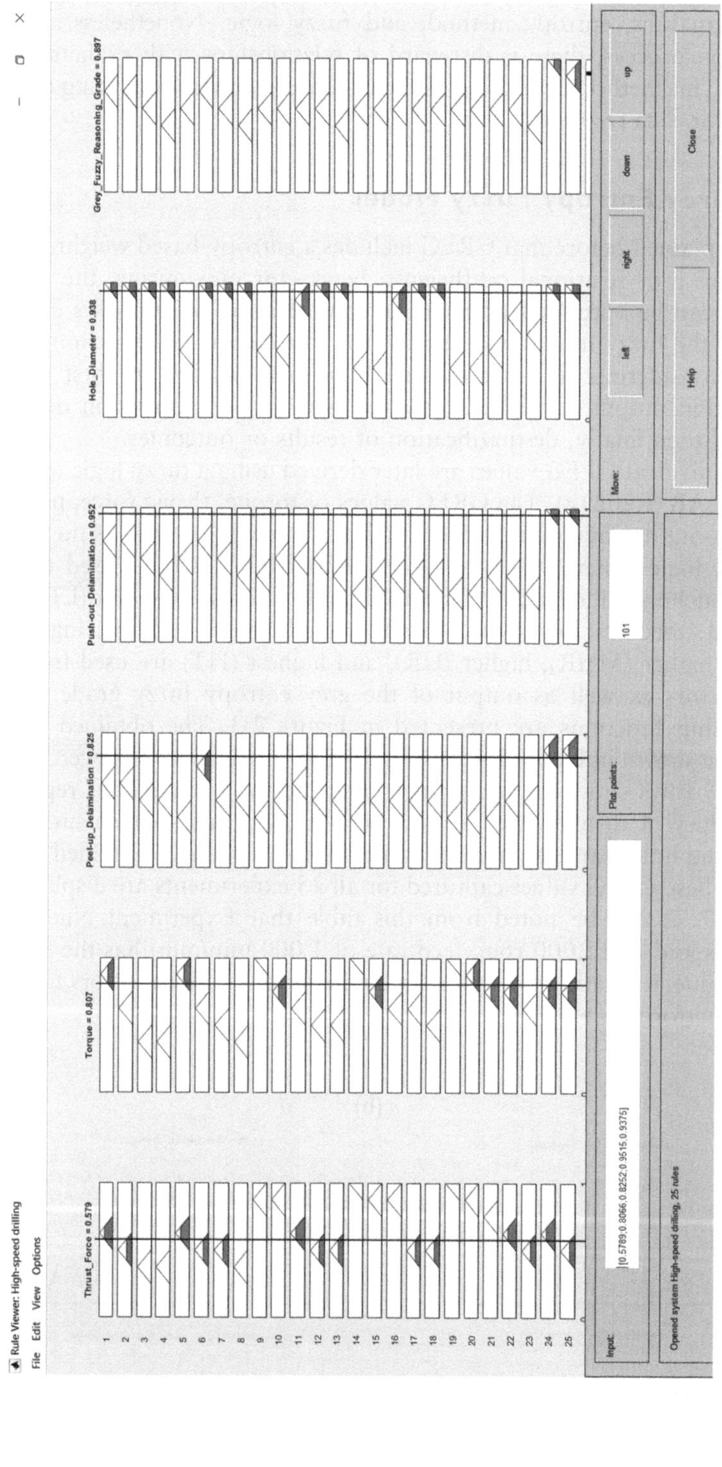


*Figure 2.4* Fuzzy-logic rule viewer of Experiment Number 7 (highest GEFG).

*Table 2.7* Grey relational coefficient and grey relational grade and grey entropy fuzzy grade (GEFG)

| *Experiment no.* | *Grey relational coefficients* | | | | | *Grey relational entropy grade* | *Grey entropy fuzzy grade* |
|---|---|---|---|---|---|---|---|
| | *Thrust force (0.20)* | *Torque (0.15)* | *Peel-up delamination (0.23)* | *Push-out delamination (0.22)* | *Hole diameter (0.20)* | | |
| 1 | 0.5946 | 0.8621 | 0.6068 | 0.7055 | 0.9783 | 0.7387 | 0.750 |
| 2 | 0.5366 | 0.7576 | 0.8972 | 0.9136 | 0.9574 | 0.8198 | 0.886 |
| 3 | 0.4583 | 0.6410 | 0.6398 | 0.7498 | 0.9783 | 0.6956 | 0.750 |
| 4 | 0.3860 | 0.4237 | 0.5073 | 0.6351 | 0.9783 | 0.5928 | 0.625 |
| 5 | 0.3333 | 0.3333 | 0.4521 | 0.5009 | 0.9783 | 0.5265 | 0.501 |
| 6 | 0.6875 | 0.8621 | 0.4344 | 0.6082 | 0.5172 | 0.6040 | 0.625 |
| 7 | 0.5789 | 0.8065 | 0.8252 | 0.9515 | 0.9375 | 0.8234 | 0.897 |
| 8 | 0.5238 | 0.6410 | 0.7232 | 0.7771 | 0.9574 | 0.7297 | 0.750 |
| 9 | 0.4583 | 0.4545 | 0.5443 | 0.6730 | 1.0000 | 0.6331 | 0.625 |
| 10 | 0.3667 | 0.3968 | 0.4585 | 0.5242 | 1.0000 | 0.5536 | 0.531 |
| 11 | 0.8462 | 0.9259 | 0.3874 | 0.4531 | 0.4369 | 0.5843 | 0.625 |
| 12 | 0.7586 | 0.8065 | 0.5495 | 0.6274 | 0.5556 | 0.6482 | 0.625 |
| 13 | 0.5946 | 0.6757 | 0.5826 | 0.6673 | 0.8491 | 0.6709 | 0.625 |
| 14 | 0.5116 | 0.5319 | 0.5123 | 0.5830 | 0.9783 | 0.6239 | 0.579 |
| 15 | 0.4314 | 0.4545 | 0.4419 | 0.4777 | 0.9000 | 0.5412 | 0.562 |
| 16 | 0.9565 | 1.0000 | 0.3723 | 0.3543 | 0.3462 | 0.5741 | 0.625 |
| 17 | 0.8462 | 0.8065 | 0.4565 | 0.5020 | 0.4128 | 0.5882 | 0.625 |
| 18 | 0.7097 | 0.6757 | 0.5213 | 0.5588 | 0.9000 | 0.6661 | 0.625 |
| 19 | 0.5641 | 0.6410 | 0.4962 | 0.5598 | 0.9375 | 0.6338 | 0.625 |
| 20 | 0.4889 | 0.5102 | 0.4517 | 0.4573 | 0.9184 | 0.5625 | 0.562 |
| 21 | 1.0000 | 1.0000 | 0.3333 | 0.3333 | 0.3333 | 0.5667 | 0.625 |
| 22 | 0.8462 | 0.8621 | 0.3763 | 0.4442 | 0.3782 | 0.5585 | 0.625 |
| 23 | 0.7586 | 0.7143 | 0.4110 | 0.4592 | 0.5056 | 0.5555 | 0.625 |
| 24 | 0.5789 | 0.7143 | 0.5121 | 0.5190 | 0.6818 | 0.5912 | 0.625 |
| 25 | 0.5238 | 0.6098 | 0.4655 | 0.4419 | 0.6618 | 0.5328 | 0.500 |

## 2.6 OPTIMISATION USING ENTROPY-BASED WEIGHT INTEGRATED MULTI-VARIATE LOSS FUNCTION

Artiles-Leon [26] was the first who proposed multi-variate loss function for quality evaluation of a product. In the majority of the instances, customer wants an ideal or targeted value for any particular quality characteristics of a product. Because of this, he presented a quality loss function for ideal value. Since loss function has no dimension, hence, loss values for various quality characteristics $Y_1$, $Y_2$, ..., $Y_n$, functions of attribute variables X's may be summed up for the calculation of total loss value, which is given in Equation (2.12).

$$TSLoss(Y_1, Y_2, \ldots, Y_n, X, T) = 4\sum_{i=1}^{n}\left(\frac{Y_i(X) - t_i}{USL_i - LSL_i}\right)^2 \tag{2.12}$$

USL = Upper specified limit
LSL = Lower specified limit

The above equations consider only nominal values for the determination of loss value. Hence, this is not applicable for other types of quality characteristics, like larger is the best (L-Type) or smaller is the best (S-Type). For this consideration, it became necessary to modify the equation so that it can be applicable to all the industrial problems. Ma and Zhao [27] modified this function, which can be suitable for real-world problems. This improved multi-variate loss function could be used for selection of the best alternative in multi-criteria decision-making problems as well as parametric optimization of processes having multiple responses.

For larger-the-better type characteristic $Y_i$,

$Y_{L_i}$ is the lowest acceptable value

$Y_{U_i}$ is the highest value above which there will not be furthermore development in the performance of the product. Hence, the loss function for L-type characteristics is given in Equation (2.13)

$$L(Y_i(X), X, Y_{U_i}) = \left(\frac{Y_i(X) - Y_{U_i}}{Y_{U_i} - Y_{L_i}}\right)^2 \tag{2.13}$$

For smaller-the-better type, the loss function is given in Equation (2.14).

$$L(Y_i(X), X, Y_{L_i}) = \left(\frac{Y_i(X) - Y_{L_i}}{Y_{U_i} - Y_{L_i}}\right)^2 \tag{2.14}$$

After the modification of Artiles-Leon's function considering different types of quality attributes, all loss functions are added to obtain an imprecise multi-variate loss function, represented in Equation (2.15).

$$L(Y(X), X) = \sum\nolimits_{i \in N} 4\left(\frac{Y_i(X) - t_i}{USL_i - LSL_i}\right)^2 + \sum\nolimits_{j \in L}\left(\frac{Y_j(X) - Y_{U_j}}{Y_{U_j} - Y_{L_j}}\right)^2 + \sum\nolimits_{l \in S}\left(\frac{Y_l(X) - Y_{L_l}}{Y_{U_l} - Y_{L_l}}\right)^2 \tag{2.15}$$

Here, L, S, and N are the numerals of L-, S- and N-type quality attributes.

Babu and colleagues [28,29] modified the Equation (2.15) by including weights in the calculation of total loss value as all the responses are not equally important for the real-world problems, Weights for the individual responses can be given either by using a subjective method or an objective method. In the present study, an objective method and entropy method are used for the calculation of weightages. Detailed calculations of weights by this method are already presented in earlier sections; the same weights are used here also.

The steps followed are:

1. Determining quality features such as nominal or intended value is best, greater is best, and smaller is best (L-, N-, and S-type).
2. Determination upper specified limit, lower specified limit, and targeted value.
3. Calculation of individual loss value for each response considering the weightage for each response.
4. Calculation of total loss by adding the individual loss values to obtain a single response.
5. Determination optimum combination of machining parameters based on the minimum total loss value.

The same experimental results as shown in Table 2.2 are chosen for the decision matrix. Weights, as calculated by an entropy method, as shown in Table 2.6, were used in the calculation of total loss value.

As the thrust force and torque should be minimum to minimise the defects during drilling, smaller the better type, Equation (2.14) and delamination (peel-up and push-out) nominal value is one, and hole diameter target or nominal value is 5, nominal type Equation (2.12) and to calculate the total loss, Equation (2.15) is used. The upper and lower specified limits are chosen from the highest and lowest values of the corresponding quality characteristics and nominal or targeted values for delamination (peel-up and push-out) and hole diameter are chosen as 1 and 5, respectively. Table shows the chosen values of upper specified and lower specified and targeted values for the responses. Maximum and minimum values of the responses were used as upper specified and lower specified limits, respectively. Lower specified limit, upper specified limit, and nominal values chosen are shown in Table 2.8. Calculated individual loss and total loss value are represented in Table 2.8.

*Table 2.8* $Y_L$, $Y_U$, and $t_i$, values

| *Response* | *Thrust force* | *Torque* | *Peel-up delamination* | *Push-out delamination* | *Hole diameter* |
|---|---|---|---|---|---|
| $Y_L$ | 16 | 0.03 | 1.112 | 1.113 | 5.00 |
| $Y_U$ | 60 | 0.28 | 2.954 | 3.391 | 5.90 |
| $t_i$ | — | — | 1 | 1 | 5 |

From Table 2.8, it can be observed that Experiment No. 7 has the lowest total loss value, 0.0417. Hence, the combination of spindle rotation of 12,000 rpm and feed of 1,000 mm/min are the optimum values of machining conditions for optimum drilling performance. The same combination of machining conditions is obtained by using both grey entropy reasoning grade and grey entropy fuzzy grade methods. In order to compare the optimisation methods used in this study, the Spearman correlation coefficient is calculated. The calculated Spearman correlation coefficient between the two approaches used in this study is 0.81, which is greater than 0.7, indicates there is strong positive correlation between the methods used for optimisation in this study Table 2.9.

*Table 2.9* Individual loss and total loss value

| *Experiment no.* | *Individual loss value* | | | | | *Total loss value* | *Rank* |
|---|---|---|---|---|---|---|---|
| | *Thrust force* | *Torque* | *Peel-up delamination* | *Push-out delamination* | *Hole diameter* | | |
| 1 | 0.1162 | 0.0064 | 0.4724 | 0.1919 | 0.0005 | 0.1752 | 4 |
| 2 | 0.1865 | 0.0256 | 0.0148 | 0.0098 | 0.0020 | 0.0471 | 2 |
| 3 | 0.3492 | 0.0784 | 0.3566 | 0.1227 | 0.0005 | 0.1907 | 5 |
| 4 | 0.6327 | 0.4624 | 1.0617 | 0.3638 | 0.0005 | 0.5202 | 13 |
| 5 | 1.0000 | 1.0000 | 1.6527 | 1.0934 | 0.0005 | 0.9708 | 19 |
| 6 | 0.0517 | 0.0064 | 1.9075 | 0.4570 | 0.8711 | 0.7248 | 14 |
| 7 | 0.1322 | 0.0144 | 0.0505 | 0.0029 | 0.0044 | 0.0417 | 1 |
| 8 | 0.2066 | 0.0784 | 0.1649 | 0.0907 | 0.0020 | 0.1114 | 3 |
| 9 | 0.3492 | 0.3600 | 0.7888 | 0.2602 | 0.0000 | 0.3625 | 7 |
| 10 | 0.7459 | 0.5776 | 1.5700 | 0.9074 | 0.0000 | 0.7965 | 15 |
| 11 | 0.0083 | 0.0016 | 2.8141 | 1.6050 | 1.6612 | 1.3345 | 22 |
| 12 | 0.0253 | 0.0144 | 0.7564 | 0.3886 | 0.6400 | 0.3947 | 9 |
| 13 | 0.1162 | 0.0576 | 0.5777 | 0.2738 | 0.0316 | 0.2313 | 6 |
| 14 | 0.2278 | 0.1936 | 1.0196 | 0.5635 | 0.0005 | 0.4332 | 10 |
| 15 | 0.4344 | 0.3600 | 1.7952 | 1.3168 | 0.0123 | 0.8459 | 18 |
| 16 | 0.0005 | 0.0000 | 3.1979 | 3.6599 | 3.5679 | 2.2544 | 24 |
| 17 | 0.0083 | 0.0144 | 1.5946 | 1.0842 | 2.0227 | 1.0136 | 20 |
| 18 | 0.0418 | 0.0576 | 0.9486 | 0.6869 | 0.0123 | 0.3888 | 8 |
| 19 | 0.1493 | 0.0784 | 1.1601 | 0.6811 | 0.0044 | 0.4592 | 11 |
| 20 | 0.2732 | 0.2304 | 1.6583 | 1.5521 | 0.0079 | 0.8136 | 16 |
| 21 | 0.0000 | 0.0000 | 4.5012 | 4.4067 | 4.0000 | 2.8047 | 25 |
| 22 | 0.0083 | 0.0064 | 3.0901 | 1.7251 | 2.7042 | 1.6337 | 23 |
| 23 | 0.0253 | 0.0400 | 2.3107 | 1.5281 | 0.9560 | 1.0699 | 21 |
| 24 | 0.1322 | 0.0400 | 1.0218 | 0.9463 | 0.2178 | 0.5192 | 12 |
| 25 | 0.2066 | 0.1024 | 1.4841 | 1.7575 | 0.2612 | 0.8369 | 17 |

## 2.7 CONCLUSIONS

This chapter presents multi-response optimisation using two different approaches, entropy-based weight integrated grey relational analysis with fuzzy logics further refinement by using genetic algorithms and entropy-based weight integrated multi-variate loss function. Input parameters in this study are spindle speed and feed and responses are thrust force, torque, peel-up and push-out delamination factors, and hole diameter. Based on obtained results from both approaches the following conclusions are drawn:

- As spindle rotational speed increases, delamination also increases and delamination is higher at the lower values of feeds.
- For spindle rotational speed of 12,000 rpm, the minimum delamination obtained at the feed rate 1 m/min but for higher speeds of 16,000 and 20,000 rpm, the optimum feed value is 3 m/min for minimum delamination.
- Torque and thrust force decrease with an increase in spindle speeds because of greater heat generation during drilling with high speeds and subsequent softening of the matrix of the composite.
- Torque and thrust force both increase with the feed rate as higher feed rates need grater forces and torque required for drilling.
- Hole diameter increases with an increase in spindle rotational speed and decreases with an increase in feed values because thermo-mechanical behaviour of composite material at higher spindle rotational speeds with lower feeds.
- Experiment Number 7 shows the maximum value of GREG value, 0.8234, is the optimum combination of input parameters (spindle rotational speed of 12,000 rpm and feed of 1,000 mm/min) for multi-response optimisation.
- Optimisation results of entropy-based weight integrated multi-variate loss function shows the total loss value is low i.e., 0.0417 for Experiment Number 7, indicating the optimum combination of machining conditions.
- Both optimisation approaches gave the same machining conditions for multi-response optimisation.
- Spearman correlation coefficient calculated is 0.81 and that indicates there is a strong positive correlation between the approaches used for optimisation.

## REFERENCES

1. J. H. Lee, E. G. Barathi Dassan, M. S. Zainol Abidin, et al. Tensile and compressive properties of glass fiber-reinforced polymer hybrid composite with eggshell powder. *Arab J Sci Eng* 45 (2020) 5783–5791.

2. U. Reisgen, A. Schiebahn, J. Lotte, C. Hopmann, D. Schneider, and J. Neuhaus. Innovative joining technology for the production of hybrid components from FRP and metals. *J Mater Process Technol.* 282 (2020) 116674.
3. S. Bayraktar, and Y. Turgut. Determination of delamination in drilling of carbon fiber reinforced carbon matrix composites/Al 6013-T651 stacks. *Measurement* 154 (2020) 107493.
4. F. Masoud, S. M. Sapuan, M. K. A. MohdAriffin, Y. Nukman, and E. Bayraktar. Cutting processes of natural fiber-reinforced polymer composites. *Polymers.* 12(6) (2020) 1332.
5. P. K. Kopparthi, V. R. Kundavarapu, V. R. Kaki, and B. R. Pathakokila. Modeling and multi response optimization of mechanical properties for E-glass/polyester composite using Taguchi-grey relational analysis. *Proc Inst Mech Eng Part L J Eng Process.* 235(2) (2021) 342–350.
6. J. Babu, T. Sunny, N. A. Paul, K. P. Mohan, J. Philip, and J. P. Davim. Assessment of delamination in composite materials: A review. *Proc Inst Mech Eng Part B J Eng Manuf.* 230(11) (2016) 1990–2003.
7. J. Babu, J. Philip, T. Zacharia, and J. P. Davim. Delamination in composite materials: measurement, assessment and prediction. In: Davim JP. *Machinability of Fibre Reinforced Plastics*. Berlin: De Gruyter (2015) 139–162.
8. A. S. Sobri, D. Whitehead, M. Mohamed, J. J. Mohamed, M. H. M. Amini, A. Hermawan, and M. N. Norizan. Augmentation of the delamination factor in drilling of carbon fibre-reinforced polymer composites (CFRP). *Polymers* 12 (2020) 2461.
9. U. A. Khashaba, M. S. Abd-Elwahed, K. I. Ahmed, I. Najjar, A. Melaibari, and M. A. Eltaher. Analysis of the machinability of GFRE composites in drilling processes. *Steel Compos Struct* 36 (2020) 417–426.
10. U. A. Khashaba, and A. A. El-Keran. Drilling analysis of thin woven glass-fiber reinforced epoxy composites. *J Mater Process Technol.* 249 (2017) 415–425.
11. A. Formisano, I. Papa, V. Lopresto, and A. Langella. Influence of the manufacturing technology on impact and flexural properties of GF/PP commingled twill fabric laminates. *J Mater Process Technol.* 274 (2019) 116275
12. A. T. Erturk, F. Vatansever, E. Yarar, E. A. Guven, and T. Sinmazcelik. Effects of cutting temperature and process optimization in drilling of GFRP composites. *J Compos Mater.* 55(2) (2021) 235–249.
13. K. Ramesha, N. Santhosh, K. Kiran, et al. Effect of the process parameters on machining of GFRP composites for different conditions of abrasive water suspension jet machining. *Arab J Sci Eng* 44 (2021) 7933–7943.
14. W. H. Kao. Tribological prosperities and high-speed drilling application of MoS2–Cr. *Wear* 258 (2005) 812–825.
15. S. C. Lin, and I. K. Chen. Drilling carbon fiber reinforced material at high-speed drilling. *Wear* 194 (1996) 156–162.
16. C. J. Rubio, A. M. Abrao, P. E. Faria, A. E. Correia, and J. P. Davim. Effect of high speed in drilling of glass fiber reinforced plastic; Evaluation of delamination factor. *Int J Mach Tools Manuf.* 48 (2008) 715–720.
17. S. Rawat, and H. Atta. Charecterization of dry high-speed drilling process of woven composites using machinability maps approach. CIRP *Ann Manuf Technol.* 58 (2009) 105–108.

18. V. Krishnaraj, A. P. Ramanathan, N. Elanghovan, M. S. Kumar, R. Zitoune, and J. P. Davim. Optimization of machining parameters at high-speed drilling of carbon fiber reinforced plastic (CFRP) laminates. *Compos Part B*. 43 (2012) 1791–1799.
19. J. Babu, N. A. Paul, K. P. Mohan, J. Philip, and J. P. Davim. Examination and modification of equivalent delamination factor for assessment of high-speed drilling. *J Mech Sci Technol.* 30(11) (2016) 5159–5165.
20. J. Babu, N. A. Paul, P. S. Abraham, B. N. Anoop, J. Philip, and J. P. Davim. Development of comprehensive delamination factor and it's assessment of high-speed drilling. *Proc Inst Mech Eng Part B J Eng Manuf.* 232(12) (2018) 2109–2121.
21. C. A. Abbas, C. Huang, J. Wang, et al. Machinability investigations on high-speed drilling of aluminium reinforced with silicon metal matrix composites. *Int J Adv Manuf Technol* 108 (2020) 1601–1611.
22. L. Zhang, S. Wang, W. Qiao, et al. High-speed milling of CFRP composites: a progressive damage model of cutting force. *Int J Adv Manuf Technol.* 106 (2020) 1005–1015.
23. Z. H. Zou, Y. Yun, and J. N. Sun. Entropy method for determination of weight of evaluating indicators in fuzzy synthetic evaluation for water quality assessment. *J Environ Sci* 18 (2006) 1020–1023.
24. P. Sharma, S. Singh, and D. R. Mishra. Electrical discharge machining of AISI 329 stainless steel using copper and brass rotary tubular electrode. *Procedia Mater Sci* 5 (2014) 1771–1780.
25. J. Kumar. Prediction of surface roughness in wire electric discharge machining (WEDM) process based on response surface methodology. *Int J Eng Technol* 2 (2012) 708–712.
26. N. Artiles-Leon. A pragmatic approach to multi-response problems using loss functions. *Qual Eng.* 9(2) (1996) 213–220.
27. Y. Ma, and F. Zhao. An improved multivariate loss function approach to optimization. *J Syst Sci Syst Eng.* 13(3) (2004) 318–325.
28. B. G. A. Mathew, J. Babu, J. Thanikachalam, and S. S. Bose. Improvement of surface finish and reduction of tool wear during hard turning of AISI D3 using Magnetorheological damper. *J Sci Ind Res India.* 77(1) (2018) 35–40.
29. J. Babu, D. Kumar, A. James, L. Paul, and S. Chakraborty. Applications of multivariate loss function and distance function approaches for material selection. *AIP Conf Proc.* 2273 (2020) 050015. 10.1063/5.0024554

Chapter 3

# Implementation of Modern Meta-Heuristic Algorithms for Optimizing Machinability in Dry CNC Finish-Turning of AISI H13 Die Steel Under Annealed and Hardened States

*Nikolaos A. Fountas[1], Ioannis Papantoniou[2], John Kechagias[3], Dimitrios E. Manolakos[2], and Nikolaos M. Vaxevanidis[1]*

[1]Laboratory of Manufacturing Processes and Machine Tools (LMProMaT), Department of Mechanical Engineering Educators, School of Pedagogical and Technological Education (ASPETE), Amarousion, Greece

[2]School of Mechanical Engineering, National Technical University of Athens (NTUA), Zografou, Greece

[3]Design and Manufacturing Laboratory (DML), University of Thessaly, Karditsa, Greece

**CONTENTS**



## 3.1 INTRODUCTION

AISI H13 alloyed steel has been widely used to manufacture several industrial parts such as forming tools, dies, extrusion dies, mandrels, etc. This wide range of applications has been feasible owing to the sustainable toughness and hardness of AISI H13 under elevated working temperatures. In most applications AISI H13 is in the range of 45 to 65 HRC (Kumar and Chauhan, 2015). A significant body of contributions has been devoted for

DOI: 10.1201/9781003258421-4

the examination of cutting force components as well as surface roughness indicators, not only for AISI H13 grade but for other engineering materials and superalloys as well (Boy et al., 2016; Hosseini et al., 2016; Patole and Kulkarni, 2017; Korkmaz et al., 2020; Vaxevanidis et al., 2020; Şahin and Esen, 2021; Arsene et al., 2021).

AISI H13 and other special engineering alloys require proper selection of cutting tool materials, especially when it comes to finish turning. Noticeable contributions in the field have reported the usage of cubic boron nitride (CBN) tools and poly crystalline diamond (PCD) tools in the form of cutting inserts. Such materials are mandatory for maintaining surface finish and accuracy. The rationale behind their selection is the fact that ordinary cutting materials do not sustain their chemical stability during the machining process; they exhibit rapid tool wear owing to high temperatures and strong adhesion (Elbestawi et al., 1997). Cutting tool selection should be based also to proper geometry according to the machining stage. It is widely known that hard-turning cutting inserts comprise 0.8 mm tool tip radius whilst those used for finish-turning have a smaller tool tip radius equal to 0.4 mm. Even though these conventional geometries have been widely applicable, they may restrict productivity or deteriorate quality owing to the restricted range for selecting feed rates. A cutting insert with a large tool tip radius will maintain surface quality, yet, it will lead to higher cutting forces and thus regenerative chatter. On the contrary, cutting inserts with smaller radii will reduce cutting force but they dramatically restrict the applicable range of feed rate selection for maintaining a good surface finish. To balance this trade-off between productivity and surface finish, wiper geometries for cutting inserts have been developed to provide an alternative to high surface finish (Abbas et al., 2020). Several research experiments have presented comparative results among conventional tool tip geometries and wiper geometries either by conducting actual machining operations (Abbas et al., 2020) or simulations using the finite element method (Ghani et al., 2008).

Undoubtedly every manufacturing process is affected by its corresponding process parameters (Ghosh et al., 2018). To determine feasible – if not advantageous – settings for process parameters, handbooks, and cutting tool catalogues are available to practitioners to select specific values from a constrained applicable range. However, such recommended ranges for setting process parameters are away from being optimal to satisfy performance metrics. In addition, new developments and novel aspects concerning modern materials, such recommendations are yet to be provided. Based on this context, artificial intelligence and soft computing techniques are continuously implemented to provide advantageous solutions to almost any manufacturing process; conventional and non-conventional. Obviously, emphasis is given to intelligent algorithms and artificial neural networks that have been widely employed to optimize process-related parameters regarding one, or more than one optimization objectives. Outeiro (2014) examined residual stresses and surface roughness in AISI H13 dry turning under different hardness states. Then, an approach involving a neural network and a genetic algorithm was used for optimizing cutting parameters. Pathak et al. (2015) optimized the turning parameters for AISI H13 by designing an L9 orthogonal array experiment and implementing a neural network. The model evaluated two objectives: the minimum surface roughness and the maximum material removal rate. Mia et al. (2019) performed optimization for hard turning parameters by using the teaching–learning-based optimization and bacterial foraging algorithms. The weighted sum method was used to transform the diverse responses into a single response.

From the review presented above, it is evident that the industrial focus involves both the experimental observation to control the process-related parameters that affect discrete performance metrics and the optimization capabilities through the application of either existing or new intelligent algorithms that follow genetic, evolutionary, and swarm-intelligence principles. Based on these concepts, this study has been attempted to implement a group of modern intelligent algorithms to optimize the parameter selection affecting two widely known machinability indicators, the main cutting force and surface roughness, for the CNC finish-turning of AISI H13 die steel using a wiper CBN cutting insert. Prior to this effort, experimental observations have been examined by implementing statistical techniques while regression equations have been generated to produce prediction models for main cutting force and surface roughness. The models have been used as objective functions for the selected algorithms and the results obtained have been rigorously compared to investigate their functionality and algorithmic evaluation modules.

## 3.2 EXPERIMENTAL

### 3.2.1 Design of Experiments

In order to examine the influence of the independent process parameters $n$ (rpm), $f$ (mm/rev), and $a$ (mm), on the responses of $Fz$ (Nt) and $Ra$ (μm) experiments CNC turning experiments were executed as per the experimental design. The central composite design (CCD) involved 20 runs with reference to the number of cutting parameters and their corresponding levels (Table 3.1).

*Table 3.1* Cutting parameters and corresponding experimental levels

| *Central composite design of experiments* | | | | | |
|---|---|---|---|---|---|
| *Parameter* | *Symbol* | *Level* | | | |
| | | *Low (−1)* | *Center (0)* | *High (1)* | *Unit* |
| *Spindle speed* | *n* | 1,500 | 1,750 | 2,000 | rpm |
| *Feed rate* | *f* | 0.050 | 0.125 | 0.200 | mm/rev |
| *Depth of cut* | *a* | 0.500 | 1.000 | 1.500 | mm |

CCD experimental design is an important approach in response surface methodology (RSM). It allows determining corner, axial, and center points of the design and therefore it can lead to more controllable solution domains for fitting a second-order regression model. However CCD approach has the drawback of involving a relatively large number of experimental runs owing to the experimental replicates. As a result, the CCD method should better be selected when the number of independent variables is low (i.e. three parameters). In this work the three cutting parameters produce a reasonable number of experimental runs. By considering uniform precision for three-factor experimentation, 8 factorial points, 6 axial points, and 6 center runs, a total of 20 runs were considered.

### 3.2.2 Materials and Equipment

The workpiece material was the commercial "*ORVAR® SUPREME*" tool steel of the known Swedish manufacturer Uddeholm®. The steel has been used in its soft-annealed condition (delivery condition 180 HB = 10 HRC) and in a hardened state with a hardness equal to 513–534 HB (53–54 HRC). Two Ø30 x 300 mm pre-machined rods comprising ten discrete zones separated by 5 mm grooves were used for the main experiments for facilitating chip removal among the cuts (Figure 3.1). The CBN wiper cutting insert used was the SECO® TNGA332S-00820-L1-C, CBN200 with the PTJNR 2525M16 insert holder (Figure 3.2) whilst machining experiments were conducted using a HAAS® TL-1 CNC turning center. The CNC turning center was equipped with a computer-aided interface (Labview® module) so as to obtain online measurements for the three cutting force components, $Fx$, $Fy$ and $Fz$ (Nt). The three-component KISTLER® dynamometer was implemented for measuring $Fz$ (Nt) (Figure 3.3).

The trends for online measurements referring to the main cutting force $Fz$, obtained during the experiments were further examined to compute the average values from raw data. Therefore, the average value from the meaningful regions (i.e. where high cutting force signals occurred), was calculated to establish the first response. Arithmetic surface roughness average $Ra$ (μm)

*Figure 3.1* Ø30 x 300 mm *ORVAR® SUPREME* bars for CNC turning experiments.

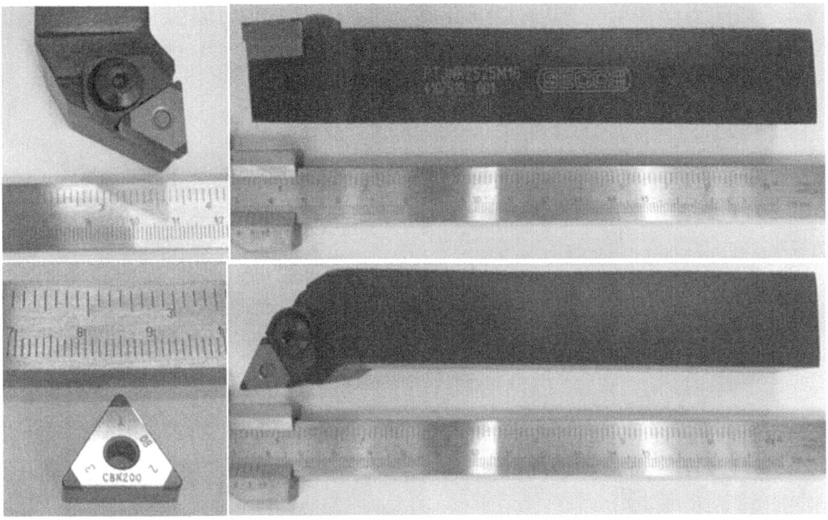

*Figure 3.2* The SECO® TNGA332S-00820-L1-C, CBN200 with the PTJNR 2525M16 insert holder.

was the second response and it was measured using a TESA® Rugorsurf® 10G portable roughness tester along with its respective software as per the *ISO-4287* standard (Figure 3.3). Each cutting zone was measured three times by rotating the workpiece at an angle of 120°. As a final surface roughness output, the average value from the three measurements was computed.

### 3.2.3 Experimental Results and Analysis

The results obtained from the experiments are presented in Table 3.2. The term *ANLD* refers to the "annealed" state of AISI H13, whereas the "HRC" indication prompts to the hardened state of the material. The asterisk "*" denotes the corrected values in the response surface experiments. Analysis of variance (ANOVA) for the results was conducted to assess the effect of machining parameters and estimate the experimental error. ANOVA for the results of main cutting force – *Fz* and arithmetic surface roughness average – *Ra* for both states of the tool steel examined are given in Tables 3.3 and 3.4.

MINITAB® R17 software was applied to statistically analyze the experimental data. The regression model selected was the full quadratic response surface model, as Equation 3.1 represents.

$$y = \beta_0 + \sum_{i=1}^{k} \beta_i x_i + \sum_{i=1}^{k} \beta_{ii} x_i x_i + \sum_{i<j} \sum \beta_{ij} x_i x_j \tag{3.1}$$

*Figure 3.3* The HAAS® TL-1 CNC turning center used for conducting the experiments; the computer-assisted apparatus based on a Labview® module for collecting the raw data for cutting force components, and the TESA® Rugorsurf® 10G portable roughness tester.

*Table 3.2* Experimental results for main cutting force (*Fz*) and surface roughness (*Ra*)

| *No.* | *n (rpm)* | *f (mm/rev)* | *a (mm)* | *Fz (N) ANLD* | *Ra (μm) ANLD* | *Fz (N) HRC* | *Ra (μm) HRC* |
|---|---|---|---|---|---|---|---|
| 1 | 1500.00 | 0.050 | 0.5000 | 109.554 | 0.754 | 142.420 | 1.276 |
| 2 | 2,000.00 | 0.050 | 0.5000 | 95.570 | 0.871 | 124.241 | 1.029 |
| 3 | 1,500.00 | 0.200 | 0.5000 | 245.129 | 9.739 | 318.668 | 1.951 |
| 4 | 2,000.00 | 0.200 | 0.5000 | 238.193 | 20.867 | 309.651 | 1.988 |
| 5 | 1,500.00 | 0.050 | 1.5000 | 206.789 | 1.939 | 268.826 | 2.785 |
| 6 | 2,000.00 | 0.050 | 1.5000 | 148.025 | 11.487 | 192.433 | 0.946 |
| 7 | 1,500.00 | 0.200 | 1.5000 | 501.935 | 15.073 | 652.515 | 3.587 |
| 8 | 2,000.00 | 0.200 | 1.5000 | 431.547 | 18.670 | 561.011 | 9.202 |
| 9 | 1,750.00 | 0.125 | 1.0000 | 256.925 | 8.535 | 334.003 | 3.941 |
| 10 | 1,750.00 | 0.125 | 1.0000 | 276.983 | 10.590 | 360.078 | 2.306 |

*(Continued)*

*Table 3.2* (continued)

| No. | *n (rpm)* | *f (mm/rev)* | *a (mm)* | *Fz (N) ANLD* | *Ra (μm) ANLD* | *Fz (N) HRC* | *Ra (μm) HRC* |
|---|---|---|---|---|---|---|---|
| 11 | 1,750.00 | 0.125 | 1.0000 | 208.133 | 2.786 | 270.573 | 2.842 |
| 12 | 1,750.00 | 0.125 | 1.0000 | 197.936 | 2.611 | 257.317 | 3.196 |
| 13 | 1,342.00* | 0.125 | 1.0000 | 202.824 | 2.128 | 263.671 | 3.798 |
| 14 | 2,158.00* | 0.125 | 1.0000 | 159.904 | 3.900 | 207.875 | 3.378 |
| 15 | 1,750.00 | 0.025* | 1.0000 | 72.099 | 6.450 | 93.729 | 5.585 |
| 16 | 1,750.00 | 0.250* | 1.0000 | 274.696 | 17.090 | 357.105 | 8.685 |
| 17 | 1,750.00 | 0.125 | 0.1800* | 18.513 | 3.117 | 24.067 | 6.303 |
| 18 | 1,750.00 | 0.125 | 1.8200* | 219.611 | 4.602 | 285.494 | 9.306 |
| 19 | 1,750.00 | 0.125 | 1.0000 | 133.241 | 4.812 | 173.213 | 9.730 |
| 20 | 1,750.00 | 0.125 | 1.0000 | 111.597 | 2.143 | 145.076 | 9.730 |

*Table 3.3* ANOVA results for *Fz* and *Ra* at the "annealed-ANLD" AISI H13 state

| *ANOVA results for main cutting force, Fz regression model* | | | | |
|---|---|---|---|---|
| | *Seq SS* | *Adj SS* | *Adj MS* | *% contribution* |
| *Model* | 133406 | 133406 | 14822.9 | 83.82 |
| *Linear* | 73010 | 73010 | 24336.8 | 49.46 |
| *Square* | 47644 | 47644 | 15881.2 | 29.22 |
| *Interactions* | 12752 | 12752 | 4250.7 | 5.14 |
| *Pure error* | 114490 | 114490 | 11449 | 16.18 |
| *Lack-of-fit (p-value)* | 0.318 | | | |
| *ANOVA results for surface roughness, Ra regression model* | | | | |
| | *Seq SS* | *Adj SS* | *Adj MS* | *% contribution* |
| *Model* | 341.553 | 341.553 | 37.950 | 85.11 |
| *Linear* | 256.122 | 256.122 | 85.374 | 63.83 |
| *Square* | 72.396 | 72.396 | 24.132 | 19.56 |
| *Interactions* | 13.035 | 13.035 | 4.345 | 1.72 |
| *Pure error* | 18.928 | 18.928 | 2.552 | 14.89 |
| *Lack-of-fit-(p-value)* | 0.269 | | | |

In Equation 3.1, Y represents the response (i.e. *Fz* (Nt) and *Ra* (μm)) whilst $x_i$ is the $i_{th}$ parameter. A result less than 0.05 for the p-value suggests that the corresponding independent variable is significant. When it comes to lack of fit, *p*-value has to be greater than 0.05. Note that an insignificant lack of fit is preferred, which means that any term excluded by the model is insignificant and thus the model fits well. Anderson–Darling normality test is used to validate the generated models' suitability referring to *Fz* (Nt) and *Ra* (μm) responses. In the Anderson–Darling test, if *p* is lower than the selected

*Table 3.4* ANOVA results for *Fz* and *Ra* at the "hardened-HRC" AISI H13 state

| *ANOVA results for main cutting force, Fz regression model* | | | | |
|---|---|---|---|---|
| | *Seq SS* | *Adj SS* | *Adj MS* | *% contribution* |
| *Model* | 225457 | 225457 | 25050.7 | 83.82 |
| *Linear* | 73010 | 73010 | 24336.8 | 59.45 |
| *Square* | 47644 | 47644 | 15881.2 | 19.22 |
| *Interactions* | 12752 | 12752 | 4250.7 | 5.14 |
| *Pure error* | 114490 | 114490 | 11449 | 16.19 |
| *Lack-of-fit (p-value)* | 0.318 | | | |

| *ANOVA results for surface roughness, Ra regression model* | | | | |
|---|---|---|---|---|
| | *Seq SS* | *Adj SS* | *Adj MS* | *% contribution* |
| *Model* | 147.202 | 147.202 | 16.357 | 89.69 |
| *Linear* | 16.083 | 16.083 | 5.361 | 8.71 |
| *Square* | 114.759 | 114.759 | 38.253 | 62.13 |
| *Interactions* | 16.360 | 16.360 | 5.453 | 18.86 |
| *Pure error* | 18.928 | 18.928 | 2.552 | 10.31 |
| *Lack-of-fit-(p-value)* | 0.123 | | | |

significance level (c.i. = 0.05), the data fails to follow a normal distribution. In this study ANOVA results for the quadratic models generated, indicate that the models are suitable for predicting *Fz* (Nt) and *Ra* (μm). The coefficient of determination ($R^2$) indicates the percentage of total variation in the response explained by the terms in the models. In this work, ANOVA indicates that all four quadratic models for the "ANLD" and "HRC" AISI H13 conditions are suitable for predicting *Fz* (Nt) and *Ra* (μm) with quite high contributions, i.e. 83.82% for both main cutting force plots; and 85.11% and 89.69% for surface roughness at the "annealed" and "hardened" conditions, respectively. *P*-values for lack of fit are both far beyond 0.05 ($p$-value > 0.25 and > 0.123); see Figure 3.4.

### 3.2.4 Statistical Analysis

With reference to $p$-value, it has been concluded that in both cases of annealed and hardened AISI H13 conditions, the main cutting force *Fz* (Nt) are mainly influenced by the linear terms followed by the square terms and interaction terms. On the contrary, when examining surface roughness *Ra* (μm) at the annealed state of AISI H13, the hierarchy of influence suggests first the linear terms, then the square terms, and finally the interaction terms, while at the hardened state square terms come first, followed by interaction effects and finally the linear effects. Individual significance of each term is calculated by a $t$-test at 95% confidence level; thus, terms

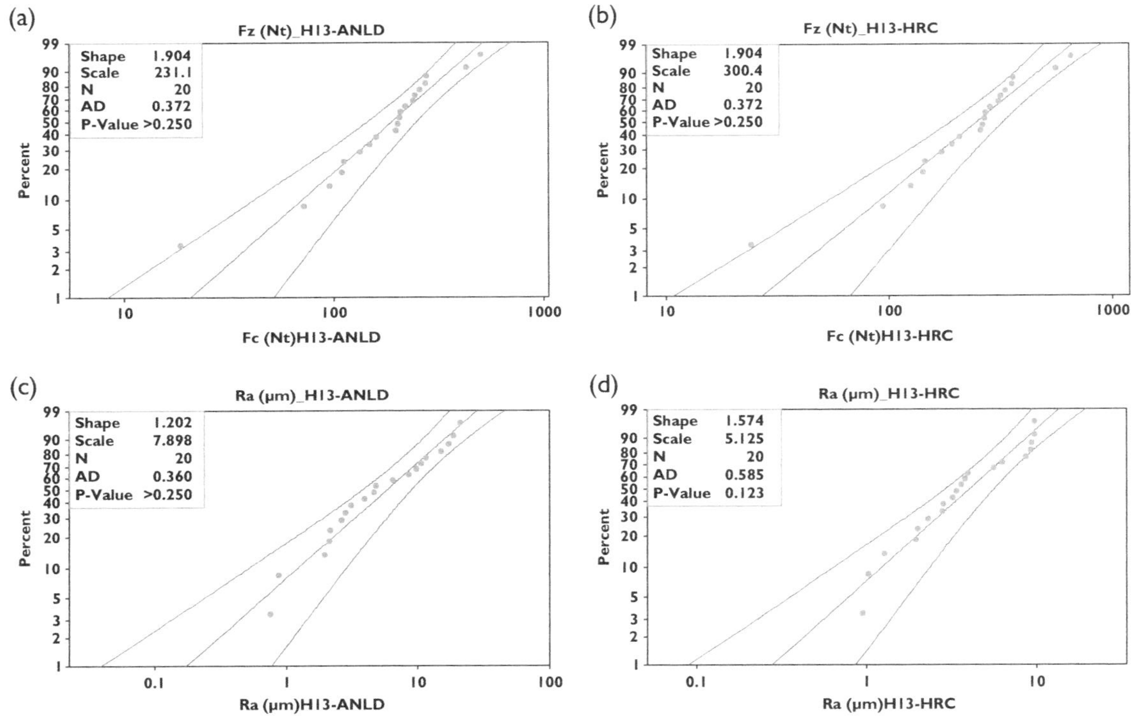


*Figure 3.4* Probability plots for regression models: (a) *Fz* for the annealed condition of AISI H13, (b) *Fz* for the hardened condition of AISI H13, (c) *Ra* for the annealed condition of AISI H13, (d) *Ra* for the hardened condition of AISI H13.

having *p*-value less than 0.05 are significant. The coefficient of determination ($R^2$) indicates the percentage of total variation in the response explained by the terms in the models and has been found equal to 83.82% for main cutting force *Fz* (Nt) in both AISI H13 conditions and 85.11% and 89.69% for surface roughness *Ra* (μm) in annealed and hardened AISI H13 conditions, respectively.

Contour plots are an alternative depiction of three-dimensional surfaces on a two-dimensional illustration. They involve two predictors-variables on X and Y axes whilst the response variable appeared in Z axis as a contour. Corresponding contour plots for the responses of main cutting force and surface roughness were generated to show how their values change as functions of different pairs of independent variables. Figure 3.5 shows the resulting changes in main cutting force and surface roughness when altering the influential process parameters regarding the response under examination, i.e., feed rate and depth of cut for cutting force and feed rate and spindle speed for surface roughness (i.e. $Fc = f(\mathrm{f}, \mathrm{a})$ and $Ra = f(\mathrm{n}, \mathrm{f})$). It is evident that feed rate *f* (mm/rev) and depth of cut *a* (mm) have the largest effect on the response of main cutting force *Fz* (Nt) referring to both material conditions. Main cutting force is maintained to low levels if moderate feeds are to be applied in combination to low-to-moderate depths of cut. The main cutting force reaches its highest value for the highest levels for feed rate and depth of cut as expected. In general surface roughness is maintained at low-to-moderate spindle speeds, with low-to-moderate feeds, while higher values for a spindle may be used only in combination to low feeds to avoid excessive tool wear, mainly referring to the hardened state.

## 3.3 MULTI-OBJECTIVE OPTIMIZATION USING MODERN META-HEURISTICS

For both conditions of the AISI H13 die steel two bi-objective optimization problems have been formulated and solved in this section, using three modern meta-heuristics, namely multi-objective greywolf algorithm, MOGWO (Mirjalili et al., 2016), multi-verse optimization algorithm MOMVO, (Mirjalili et al., 2017), and multi-objective antlion algorithm, MOALO (Mirjalili et al., 2017). Main cutting force *Fz* and surface roughness *Ra*, are the two optimization objectives with respect to the three cutting conditions of spindle speed *n* (rpm), feed rate *f* (mm/rev), and depth of cut *a* (mm). The solution domain has been created by adhering to the same parameter low-high levels whilst each candidate solution is a vector corresponding to the values of three machining parameters within their predefined ranges. The two problems have been solved using the recommended settings for algorithm-specific parameters, with reference 20 individuals and 1,000 generations as the major algorithmic settings for running the simulations in MATLAB® 2014b. For all three algorithms, 50 non-dominated solutions

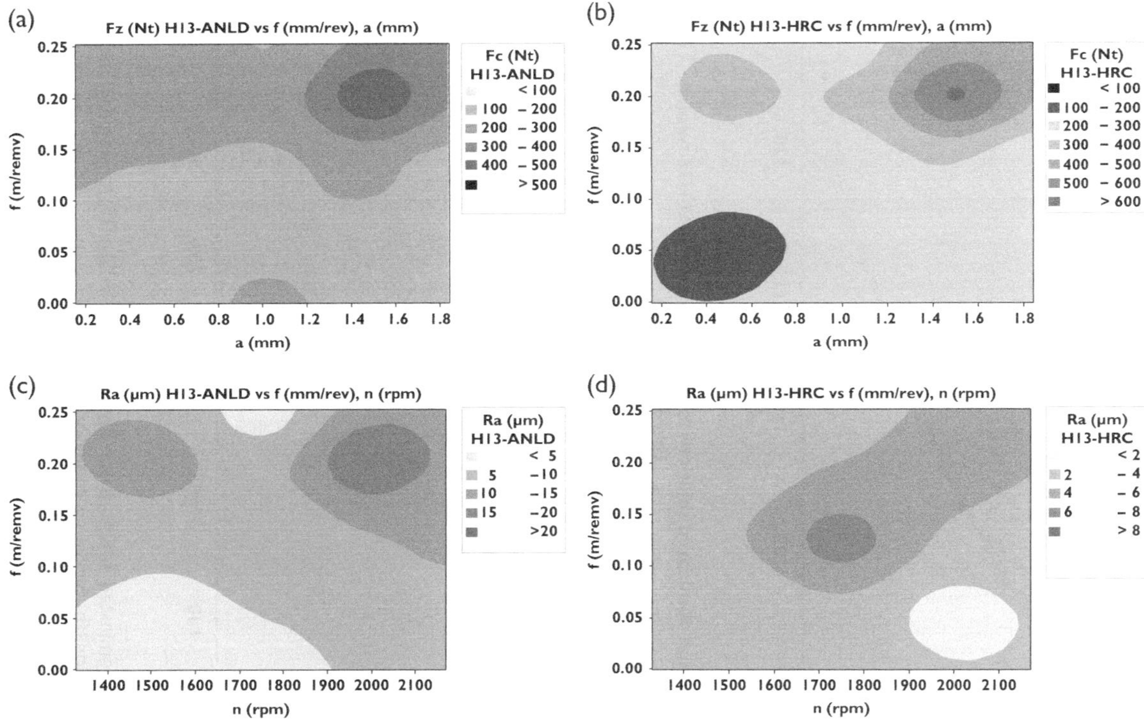


*Figure 3.5* Contour plots for: (a) *Fz* for the annealed condition of AISI H13, (b) *Fz* for the hardened condition of AISI H13, (c) *Ra* for the annealed condition of AISI H13, (d) *Ra* for the hardened condition of AISI H13.

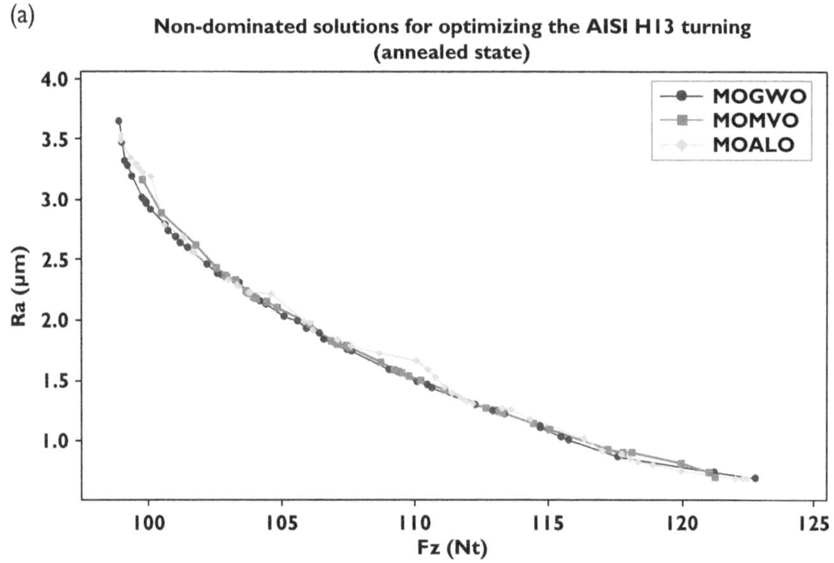


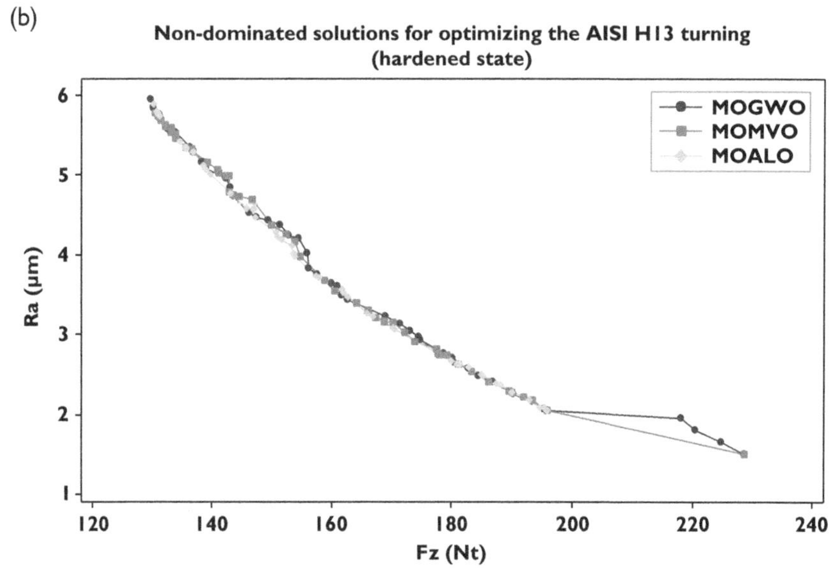


*Figure 3.6* Non-dominated solutions for optimizing the AISI H13 turning: (a) annealed condition, (b) hardened condition.

were kept in the output archive. Figure 3.6 depicts the best non-dominated solutions set observed by conducting several independent runs to examine variability in optimal solutions. It can be observed that the non-dominated solutions set corresponding to MOGWO's simulation behavior, is more uniform compared to the other two sets (MOMVO, MOALO) in the case of the AISI H13 annealed state, while the majority of points are in the region where both objectives are satisfied. In the hardened case of AISI H13, the results appear to be more complex with the impression that MOALO has managed to outperform the non-dominated solutions of MOGWO and MOMVO. However, its spread is narrower than those corresponding to MOGWO and MOMVO referring to the X-axis assigned to *Fz*. These solutions clearly add to cutting force but ease surface roughness to different percentages. It is the job of the end user to decide which of these solutions should be implemented regarding the production needs and priorities in terms of machining objectives.

## 3.4 CONCLUSIONS

This work has studied the effect of cutting parameters spindle speed *n*, feed rate *f*, and depth of cut *a* on main cutting force *Fz*, and surface roughness *Ra*. The research refers to two conditions of the AISI H13 die steel; soft-annealed (10 HRC) and hardened (53–54 HRC). Experimental results were obtained through the utilization of response surface methodology by establishing a central composite design. The statistical analysis and interpretation of the respective results was carried out by performing analysis of variance and regression modeling. Normal probability and contour plots were exploited to study the effects of parameters on the responses. The experimental outputs were further used for building regressions models that served as objective functions for optimizing the objectives of cutting force and surface roughness using three modern intelligent algorithms namely MOGWO, MOMVO, and MOALO. The findings of the study are summarized as follows:

- When finish-turning the AISI H13 die steel at its hardened condition, main cutting force is approximately 30% larger the one corresponding to the soft-annealed state. Yet, surface roughness is reduced providing superior surface finish.
- According to the analysis of variance, the hierarchy of effects of cutting parameters in terms of cutting force suggests the linear terms, the square terms and finally the interaction terms regardless of the AISI H13 condition. However, when it comes to surface roughness, this hierarchy changes to linear, followed by square, followed by interaction terms for annealed state, and square, followed by interaction, followed by linear terms for the hardened state.
- Depth of cut and feed rate are influential cutting parameters for main cutting force, whilst spindle speed and feed rate are influential cutting parameters for surface roughness, regardless of the material condition. Surface roughness alters its experimental trend from one condition to another with quite high complexity; thus, suggesting a difficulty in correlating the independent machining parameters to surface roughness.

- There is no clear superiority in multi-objective intelligent algorithms when implemented to machining optimization problems. Therefore, algorithms should be tested by conducting several evaluations and examining their statistical outputs to gain a clear understanding for their performance. However, final selections for the settings of advantageous machining parameters to facilitate all objective under question, should be based on the requirements corresponding to production and shop floor's prerequisites, since no unique solution can be deemed superior to other when it comes to multi-objective optimization.

## ACKNOWLEDGMENTS

This work was funded by the "Research & Management committee – E.L.K.E. ASPETE" under the auspices of the Research program "Parametric Analysis & Machining Parameters Optimization of Special Purposes Steels" for the initiative "Research strength in ASPETE 2018–2020"-Ref.No.: 80156.

## REFERENCES

Abbas, A. T., El Rayes, M. M., Luqman, M., Naeim, N., Hegab, H., & Elkaseer, A. (2020). On the Assessment of Surface Quality and Productivity Aspects in Precision Hard Turning of AISI 4340 Steel Alloy: Relative Performance of Wiper vs. Conventional Inserts. *Materials*, *20*, 2036. 10.3390/ma13092036

Arsene, B., Gheorghe, C., Sarbu, F. A., Barbu, M., Cioca, L.-I., & Calefariu, G. (2021). MQL-Assisted Hard Turning of AISI D2 Steel with Corn Oil: Analysis of Surface Roughness, Tool Wear, and Manufacturing Costs. *Metals*, *11*, 2058. 10.3390/met11122058

Boy, M., Yaşar, N., & Çiftçi, İ. (2016). Experimental Investigation and Modelling of Surface Roughness and Resultant Cutting Force in Hard Turning of AISI H13 Steel. *IOP Conference Series Materials Science and Engineering*, *161*(1), 012039. 10.1088/1757-899X/161/1/012039

Elbestawi, M. A., Chen, L., Becze, C. E., & El-Wardany, T. I. (1997). High-speed Milling of Dies and Molds in their Hardened State. *CIRP Annals*, *46*, 57–62.

Ghani, M. U., Abukhashim, N. A., & Sheikh, M. A. (2008). An Investigation of Heat Partition and Tool Wear in Hard Turning of H13 Tool Steel with CBN Cutting Tools. *International Journal of Advanced Manufacturing Technology*, *39*(9–10), 874–888.

Ghosh, P. S., Chakraborty, S., Biswas, A. R., & Mandal, N. K. (2018). Empirical Modelling and Optimization of Temperature and Machine Vibration in CNC Hard Turning. *Materials Today: Proceedings*, *5*(5), Part 2, 12394–12402. 10.1016/j.matpr.2018.02.218

Hosseini, A., Hussein, M., & Kishawy, H. A. (2016). On the Machinability of Die/Mold D2 Steel Material. *International Journal of Advanced Manufacturing Technology*, *85*(1–4), 735–740.

Korkmaz, M. E., Yaşar, N., & Günay, M. (2020). Numerical and Experimental Investigation of Cutting Forces in Turning of Nimonic 80A Superalloy. *Engineering Science and Technology, an International Journal*, *23*(3), 664–673.

Kumar, P., & Chauhan, S. R. (2015). Machinability Study on Finish Turning of AISI H13 Hot Working Die Tool Steel With Cubic Boron Nitride (CBN) Cutting Tool Inserts Using Response Surface Methodology (RSM). *Arabian Journal of Science and Technology*, *40*, 1471–1485.

Mia, M., Królczyk, G., Maruda, R., & Wojciechowski, S. (2019). Intelligent Optimization of Hard-Turning Parameters Using Evolutionary Algorithms for Smart Manufacturing. *Materials*, *12*, 879. 10.3390/ma12060879

Mirjalili, S., Jangir, P., & Saremi, S. (2017). Multi-objective Ant Lion Optimizer: A Multi-Objective Optimization Algorithm for Solving Engineering Problems. *Applied Intelligence*, *46*(1), 79–95.

Mirjalili, S., Jangir, P., Zahra, S., Mirjalili, Sh., Saremi, I., & Trivedi, N. (2017). Optimization of Problems with Multiple Objectives Using the Multi-verse Optimization Algorithm. *Knowledge-Based Systems*, *134*, 50–71.

Mirjalili, S., Saremi, S., Mirjalili, S. M., & Coelho, L. S. (2016). Multi-objective Grey Wolf Optimizer: A Novel Algorithm for Multi-criterion Optimization. *Expert Systems with Applications*, *47*(1), 106–119.

Outeiro, J. C. (2014). Surface Integrity Predictions and Optimisation of Machining Conditions in the Turning of AISI H13 Tool Steel. *International Journal of Machining and Machinability of Materials*, *15*(1–2), 122–134.

Pathak, H., Das, S., Doley, R., & Kashyap, S. (2015). Optimization of Cutting Parameters for AISI H13 Tool Steel by Taguchi Method and Artificial Neural Network. *International Journal of Materials Forming and Machining Processes*, *2*(2), 47–65.

Patole, P. B., & Kulkarni, V. V. (2017). Experimental Investigation and Optimization of Cutting Parameters with Multi Response Characteristics in MQL Turning of AISI 4340 Using Nano Fluid. *Cogent Engineering*, *4*(1), 1–14. 10.1080/23311916.2017.1303956

Şahin, E., & Esen, İ. (2021). Statistical and Experimental Investigation of Hardened AISI H11 Steel in CNC Turning with Alternative Measurement Methods. *Advances in Materials Science and Engineering*, 2021, 9480303. 10.1155/2021/9480303

Vaxevanidis, N. M., Fountas, N. A., Papantoniou, I., & Manolakos, D. E. (2020). Experimental Investigation and Regression Modelling to Improve Machinability in CNC Turning of CALMAX® Tool Steel Rods. *2020 IOP Conference Series Materials Science and Engineering*, *968*(1), 012012. 10.1088/1757-899X/968/1/012012

Chapter 4

# Multi-Response Optimization in Turning of UD-GFRP Composites Using Weighted Principal Component Analysis (WPCA)

*Meenu and Surinder Kumar*
Department of Mechanical Engineering, National Institute of Technology, Kurukshetra, Haryana, India

## CONTENTS



## 4.1 INTRODUCTION

In this research work, MRR and surface irregularity of the manufactured goods obtained through the turning process are calculated experimentally and the outcome obtained is interpreted systematically. The most important industries utilising composites are wind energy, aerospace, shipping, and building (Kazmierski, 2012). Composite material is prepared of two or more dissimilar materials and gives out properties that are not attainable through several particular material components. In this heterogeneous material, one material is the matrix and the material with a small amount is the reinforcement.

The difficulties encountered during machining of composite materials are matrix burning, pattern of fine particles like chips, and rapid tool wear (Wang and Zhang, 2003 & Davim, 2009). Considering these shortcomings, the conventional machining process still finds a wider acceptance for machining composite material. Various variables such as the workpiece material, the

DOI: 10.1201/9781003258421-5

wounding tool material, the inflexibility of the machine and the position, and management have to be considered.

Presented are various studies on the machining of GFRPs. The machining of GFRP composites is disparate as of conservative materials and it necessitates strange consideration on wear resistance of the tool. To find surface irregularity and dimensional property requires an optimization technique to locate the best achievable cutting parameters (Palanikumar, 2008). Eyup Bagci and Birhan Isik (2006) investigate the process of machining of UD-GFRP composites. The maximum test errors in predicting surface roughness were 6.30% and 6.36% from the ANN model and RSM, respectively. Isik (2008) obtained the results of the machining of UD-GFRP and the most favorable cutting parameters were recommended to acquire a desired surface quality. Isik (2009) used weighting techniques for the rotary of UD-GFRP composites with a cemented carbide instrument. Meenu Gupta and Surinder Kumar (2015) studied the optimization of surface irregularity and MRR in rotating a UD-GFRP rod through a PCD cutting tool with the principal component analysis technique. It was established that feed is a superfluous major factor, followed by DOC and speed.

Parida et al. (2014) investigated the compressive properties of epoxy resin when combined with graphite and powder clay under different conditions. The effect of cutting speed during machining of GFRP was studied. Khandey et al. (2017) used GA combined with GRA based on PCA for the machining of aluminum-based MMC with 25% SiC particulates using a turning process. Surinder Kumar et al. (2019) presented a fuzzy logic optimization technique to optimize the turning of a UD-GFRP workpiece by using a PCD cutting tool. The DOC was established as the most essential factor. (Yang, 2012) compared the performance of genetic algorithm, PSO, and flower pollination applied to different optimization problems. It was found that the flower algorithm (FA) is more powerful than GA and PSO.

Moshat et al. (2010) utilized a PCA-based Taguchi method to optimize surface irregularity and MRR in a CNC vertical milling where machining aluminum plates is done utilizing CVD-coated carbide tools. The milling parameters were feed, DOC, and speed. It was found that the proposed method is capable of solving multi-faceted difficulty. Anantha kumar et al. (2013) optimized surface irregularity, MRR, and tool wear in the turning of 1040 medium carbon steel using a PCA. From the experimentation results, DOC and feed showed more superior control than speed on the combined effect. Mohanty et al. (2019) applied weighted a PCA combined with the Taguchi process during EDM of D2 steel with copper, brass, and DMLS electrodes. A tool electrode was obtained as the important factor through the ANOVA. Chintan Kayastha et al. (2013) used PCA shared with Taguchi's approach to forecast the surface irregularity and MRR in a copper workpiece. The variables such as feed, speed, DOC, and tool position were calculated for least surface irregularity and greatest MRR. The correlation between MRR and Ra was 0.251. Mishra et al. (2017) reduced the dimensionality of big record sets by using a PCA based on a vector space transform. The parametric algorithm first transformed the data to the appropriately centered polar coordinates and then computed the PCA.

In the last few years, several heuristic evolutionary optimization algorithms have been developed. Several techniques to resolve optimization issues found in the literature are artificial immune system (Farmer et al., 1986), ant colony (Dorigo et al., 2006), simulated annealing (Suman and Kumar, 2006), gravitational search algorithm (Esmat Rashedi et al., 2009), genetic algorithm (Tang et al., 1996), gray wolf algorithm (Mirjalili et al., 2014), bee algorithms (Pham et al., 2006), glow worm swarm (Krishnanand and Ghose, 2005), and cuckoo search (Gandomi et al., 2013). These algorithms have been useful to a big range of applications.

In this manuscript, steps are taken to optimize the parameters, such as DOC, feed, environment, speed, rake angle, and nose radius in orthogonal cutting of a UD-GFRP composite for the two responses (MRR and surface roughness) with WPCA. In the WPCA technique, every component was considered to thoroughly explain the variant in every response. The WPCA technique was used to explain the variant as the weight to join every principal component in order to create MPI (Das et al., 2013). The $L_{18}$ orthogonal array trial plan is used in the current revision. ANOVA is performed to find the involvement of every parameter on the performance attributes. These methodologies helped to gain excellent feasible tool geometry and cutting situations for the turning of UD-GFRP with a carbide (k10) cutting tool. Confirmatory analysis is conducted to validate the optimum process parameters.

## 4.2 EXPERIMENTAL STUDY

Experiments are carrying out on a NH22–lathe arrangement with a cooled, wet and dry cutting situation. Figure 4.1 shows the lathe setup.

### 4.2.1 Work Material, Cutting Tool, and Cutting Conditions

A workpiece material (GFRP rod) was used, consisting of UD fibers that are pulled throughout a resin bathtub into the form of a rod. The benefit of GFRP comprises (Harvey and Ansell Martin, 2000) easy handling, light weight, well-suited with resin and timber, well-suited material properties, advanced resistance to rotting, and convenient in an acidic atmosphere. Pultrusion processed unidirectional GFRP composite rods are used. The material has E-glass as 75±5% g, epoxy resin (weight 25±5%), density as 1.95–2.1gm/cc, water absorption 0.07%, tensile strength 6,500 kgf/cm$^2$, compression strength 6,000 kgf/cm$^2$, weight of rod is 2.300 kg, length of 840 mm, and diameter as 42 mm. A carbide tool (K10) is used in the experimentation. The rake angles of the carbide tool taken into consideration

*Figure 4.1* Lathe center.

are: +6°, 0°, and -6°. Two tool nose radii of 4 mm and 8 mm are taken into consideration for the study. The properties of a carbide (K10) tool used are as follows: density, 14.95 g/cm$^3$; clearance angle, 7°; hardness, 1,600 Vickers kg/mm$^2$; thermal conductivity, 95 W/mK; and front clearance, 10°. Taguchi's mix level plan is selected. The most convenient orthogonal array (OA) is $L_{18}$ orthogonal array with 17 degrees of freedom. The designed variables at various levels are planned in Table 4.1.

### 4.2.2 Response Variables

Surface roughness and MRR are response variables that are used in the study. The $R_a$ of the machined surface is evaluated with a Surf-com 130A type instrument with a transverse length of 4 mm, least count 0.01 μm, and cut-off length 0.8 mm. The outcome of the experiment for 18 test conditions with three repetitions is reported in Table 4.4. The chosen quality feature, surface

*Table 4.1* Design variables and their level

| *Parameter design* | *Levels* | | |
|---|---|---|---|
| | *L1* | *L2* | *L3* |
| G | 0.4 | 0.8 | NIL |
| H | −6 | 0 | +6 |
| I | 0.05 | 0.1 | 0.2 |
| J | 55.42 | 110.84 | 159.66 |
| K | Dry | Wet | Cooled |
| L | 0.2 | 0.8 | 1.4 |

roughness, is smaller-the-better and MRR is higher-the-better and the S/N ratio used is as given in Equations (4.1) and (4.2) (Roy, 1990).

Smaller the superlative quality:

$$S/N = -10 \text{ Log } \frac{1}{n} \sum y^2 \tag{4.1}$$

Higher the better quality:

$$S/N = -10 \text{ Log } \frac{1}{n} \sum \frac{1}{y^2} \tag{4.2}$$

Equation (4.3) is utilized for determination of MRR. The MRR, in $mm^3$/sec, is calculated from the following relation:

$$MRR = \frac{\frac{\pi}{4} Mo^2 L - \frac{\pi}{4} Mi^2 L}{L/IN} \tag{4.3}$$

where $N = \frac{V * 1000 * 60}{\Pi M}$

Where $M_0$ = initial dia. in mm, $M_i$ = final dia in mm, $L$ = length of the workpiece material to be turned (mm), M = mean diameter, N = spindle speed (rpm), $I$ = feed (mm/rev), and $T_m$ is the machining time defined as $T_m$ = L /I N.

The measured experimental values for MRR and $R_a$ are shown in Table 4.2.

## 4.3 WEIGHTED PRINCIPAL COMPONENT ANALYSIS

Su and Tong (1997) and Antony (2000) put forward a PCA to work out a multi-response problem. A PCA eliminates correlation between the response and finds out an uncorrelated value index called principal components (Datta, Nandi, Bandyo padhyay and Pal, 2009). It is basically a dimension reduction technique where uncorrelated features (i.e. principal component) are generated from the correlated features. To reduce the dimensionality, the principal component with less variance is ignored so that the dimensionality reduces from n→k, where k<n. In this investigation, a weighted principal component is used to resolve a multi-optimization problem. The explained variation is used as the weight to take into consideration all principal components. In industry, multi-response optimization is considered a quality characteristic.

### 4.3.1 Weighted Principal Components Analysis

Let $X_i = (X_i^1, X_i^2, \ldots X_i^k \ldots . X_i^n)$ for i= (1, 2 .... m)

*Table 4.2* Experimental results of $R_a$ and MRR

| | Variables | | | | | | Responses | | | | Responses | | | |
|---|---|---|---|---|---|---|---|---|---|---|---|---|---|---|
| | G | H | I | J | K | L | Raw data | | | Avg. $R_a$ | Raw data | | | Avg. MRR |
| Trial no. | Nose radius | Rake angle | Feed | Speed | Cutting environment | DOC | $R_a$ (μm) | | | | MRR ($mm^3$/sec) | | | |
| | | | | | | | T1 | T2 | T3 | | T1 | T2 | T3 | |
| 1 | 0.4 | −6° | 0.05 | 55.42 | Dry | 0.2 | 1.59 | 1.65 | 1.49 | 1.577 | 8.50 | 8.60 | 8.70 | 8.60 |
| 2 | 0.4 | −6° | 0.1 | 110.84 | Wet | 0.8 | 1.73 | 1.77 | 1.99 | 1.830 | 144.96 | 145.02 | 145.02 | 145.00 |
| 3 | 0.4 | −6° | 0.2 | 159.66 | Cooled | 1.4 | 2.77 | 4.12 | 5.13 | 4.00 | 329.98 | 330.23 | 330.23 | 330.15 |
| 4 | 0.4 | 0° | 0.05 | 55.42 | Wet | 0.8 | 2.20 | 2.18 | 2.04 | 2.140 | 36.24 | 36.24 | 36.24 | 36.24 |
| 5 | 0.4 | 0° | 0.1 | 110.84 | Cooled | 1.4 | 1.83 | 1.83 | 1.77 | 1.810 | 237.96 | 237.90 | 238.04 | 237.97 |
| 6 | 0.4 | 0° | 0.2 | 159.66 | Dry | 0.2 | 2.69 | 2.88 | 2.89 | 2.820 | 99.00 | 98.90 | 98.93 | 98.93 |
| 7 | 0.4 | +6° | 0.05 | 110.84 | Dry | 1.4 | 1.62 | 1.94 | 2.12 | 1.893 | 125.03 | 125.02 | 125.02 | 125.02 |
| 8 | 0.4 | +6° | 0.1 | 159.66 | Wet | 0.2 | 1.99 | 1.79 | 1.89 | 1.890 | 52.98 | 52.95 | 52.99 | 52.97 |
| 9 | 0.4 | +6° | 0.2 | 55.42 | Cooled | 0.8 | 2.58 | 2.94 | 2.10 | 2.540 | 144.92 | 145.02 | 144.90 | 144.95 |
| 10 | 0.8 | −6° | 0.05 | 159.66) | Cooled | 0.8 | 2.90 | 2.72 | 2.35 | 2.656 | 104.39 | 104.41 | 104.39 | 104.40 |
| 11 | 0.8 | −6° | 0.1 | 55.42 | Dry | 1.4 | 2.15 | 2.20 | 1.95 | 2.100 | 124.96 | 124.96 | 124.96 | 124.96 |
| 12 | 0.8 | −6° | 0.2 | 110.84 | Wet | 0.2 | 2.45 | 1.56 | 2.26 | 2.090 | 73.54 | 73.53 | 73.51 | 73.53 |
| 13 | 0.8 | 0° | 0.05 | 110.84 | Cooled | 0.2 | 1.77 | 1.55 | 1.89 | 1.736 | 18.39 | 18.39 | 18.38 | 18.39 |
| 14 | 0.8 | 0° | 0.1 | 159.66 | Dry | 0.8 | 3.05 | 2.41 | 2.51 | 2.656 | 197.70 | 197.06 | 197.92 | 197.56 |
| 15 | 0.8 | 0° | 0.2 | 55.42 | Wet | 1.4 | 2.61 | 1.87 | 3.38 | 2.620 | 240.94 | 241.06 | 240.92 | 240.97 |
| 16 | 0.8 | +6° | 0.05 | 159.66 | Wet | 1.4 | 2.26 | 2.69 | 1.96 | 2.303 | 170.00 | 170.09 | 170.00 | 170.03 |
| 17 | 0.8 | +6° | 0.1 | 55.42 | Cooled | 0.2 | 1.65 | 1.68 | 1.38 | 1.570 | 18.38 | 18.38 | 18.39 | 18.38 |
| 18 | 0.8 | +6° | 0.2 | 110.84 | Dry | 0.8 | 2.53 | 2.99 | 2.50 | 2.673 | 261.00 | 260.93 | 260.80 | 260.91 |
| | | | **Total** | | | | | | | **Overall Mean = 2.272** | | | | **Overall Mean = 132.72** |

*Nose radius, depth of cut is in mm, feed rate in mm/rev, cutting speed in m/min

$X_i$ represent the response for the $i^{th}$run, having n dimensions and m represents the number of experiments.

Step 1. To normalize the responses
The individual experimental response is normalized by subtracting the mean and then dividing by the standard deviation.

$$Xnew_i = (X_i^1, \dots X_i^2, \dots X_i^k \dots . X_i^n)$$

Step 2. To determine the corresponding eigen vectors and eigen value
To calculate the eigen value $\lambda_k$ where (k= (1, 2....n) and corresponding eigen vectors for all quality characteristic, a single-value decomposition technique is used.

Step 3. To calculate the principal component series for the data

$$Y_i^k = \sum_{j=1}^{n} Xnew_i^j \beta_{kj} \text{ where i = 1, 2 .... m, k = 1, 2 .... n}$$

where $Y_i^k$ is the principal component sequence of the $k^{th}$ element in the $i^{th}$ series. $Xnew_i^j$ is the normalized and centered value of the $j^{th}$ element in the $i^{th}$ run and $\beta_{kj}$ is the $j^{th}$ element of eigen vector $\beta_k$.

Step 4. Multiple performance index is calculated using the following equation:

$$MPI_i = \lambda_i Y_i^1 + \lambda_2 Y_i^2 \dots \dots \dots . \lambda_n Y_i^n \text{ For i = 1, 2 .... n.}$$

This MPI is the final response to be optimized.

### 4.3.2 Single-Value Decomposition

Step 1. Decompose an n*n matrix X into

$$X = USV^{-1}$$

Where eigen vectors of $XX^T$ are $U_i$ and $X^TX = V_i$. Both matrices have the same singular values. Eigen values (d) are given by the square of singular values S.

Step 2. Order the eigen vectors such that vectors with a higher eigen value come before the small values ($u_1 \dots \dots u_m$), where $u_1 \dots \dots u_m$ are the eigen vectors with $d_1 \geq d_2 \dots \dots \dots \geq d_m$.

Step 3. The principal components are found by multiplying a normalized sequence with an eigen vector.

Step 4. Weighted MPI is found by multiplying eigen values with corresponding PCs.

## 4.4 ANALYSIS AND EVALUATION OF EXPERIMENTAL RESULTS

Table 4.2 shows the parametric position and the raw data such as surface irregularity and MRR in which all experiments are conducted. Surface irregularity and metal removal rate records were analyzed to find out the influence of various parameters (Kumar et al., 2013). For Ra, (SB) criterion is used, whereas (HB) criterion is chosen for MRR, as given by Equations (4.1) and (4.2), respectively. The standard deviations for Ra and MRR are 0.5930 and 92.9546, respectively. The normalized value are shown in Table 4.3. Then the single-value decomposition is performed that gives significant singular values, eigen values, and eigen vectors, as shown in Table 4.4. Eigen values are obtained using a square of singular values. Table 4.5 gives the

*Table 4.3* Normalized data

| *Reference sequence* | *Surface roughness* | *Metal removal rate* |
|---|---|---|
| 1 | −1.1728 | −1.3353 |
| 2 | −0.7461 | 0.1321 |
| 3 | 2.9134 | 2.1239 |
| 4 | −0.2234 | −1.0379 |
| 5 | −0.7799 | 1.1323 |
| 6 | 0.9234 | −0.3635 |
| 7 | −0.6399 | −0.0828 |
| 8 | −0.6450 | −0.8579 |
| 9 | 0.4512 | 0.1316 |
| 10 | 0.6468 | −0.3047 |
| 11 | −0.2908 | −0.0835 |
| 12 | −0.3077 | −0.6368 |
| 13 | −0.9047 | −1.2300 |
| 14 | 0.6468 | 0.6975 |
| 15 | 0.5861 | 1.1645 |
| 16 | 0.0515 | 0.4014 |
| 17 | −1.1846 | −1.2301 |
| 18 | 0.6755 | 1.3791 |

*Table 4.4* Singular values, eigen values, and eigen vectors

| | *Singular values (S)* | *Eigen values (d)* | *Eigen vectors U* | |
|---|---|---|---|---|
| | | | *U1* | *U2* |
| 1 | 5.3864 | 29.0131 | −0.7071 | −0.7071 |
| 2 | 2.2331 | 4.9869 | −0.7071 | 0.7071 |

*Table 4.5* Principal components in all $L_{18}$ OA investigational observations

| *Reference sequence* | *PC1* | *PC2* |
|---|---|---|
| 1 | 1.1575 | −0.7309 |
| 2 | 0.4305 | 0.6173 |
| 3 | −3.5619 | −0.5582 |
| 4 | 0.8919 | −0.5760 |
| 5 | −0.2492 | 1.3521 |
| 6 | −0.3959 | −0.9100 |
| 7 | 0.5111 | 0.3939 |
| 8 | 1.0627 | −0.1506 |
| 9 | −0.4121 | −0.2260 |
| 10 | −0.2419 | −0.6728 |
| 11 | 0.2647 | 0.1466 |
| 12 | 0.6678 | −0.2327 |
| 13 | 1.5094 | −0.2300 |
| 14 | −0.9506 | 0.0359 |
| 15 | −1.2379 | 0.4090 |
| 16 | −0.3203 | 0.2474 |
| 17 | 1.7074 | −0.0321 |
| 18 | −1.4528 | 0.4975 |

principal components that are found by multiplying a normalized sequence with an eigen vector. Figure 4.2 shows the principal components and the transformed data. Figure 4.3 shows the scree plot. Weighted MPI is found by multiplying eigen values with corresponding PCs and are normalized in the

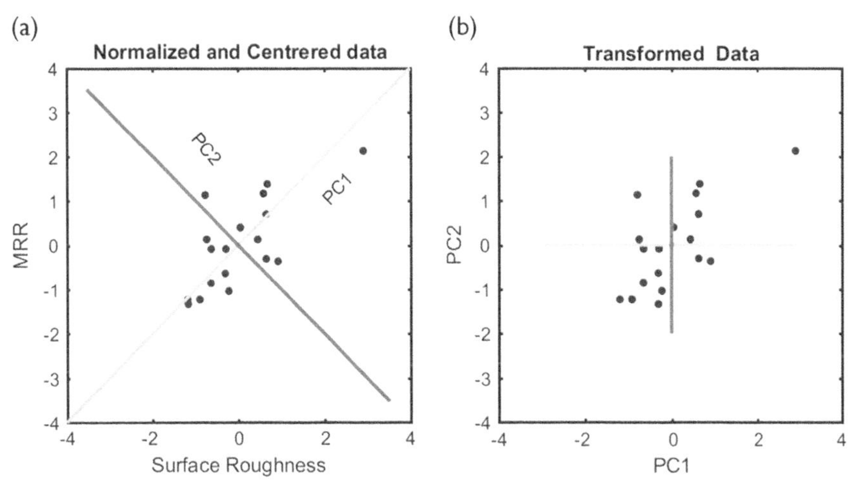


*Figure 4.2* (a) Principal components of normalized and centered data, (b) transformed data.

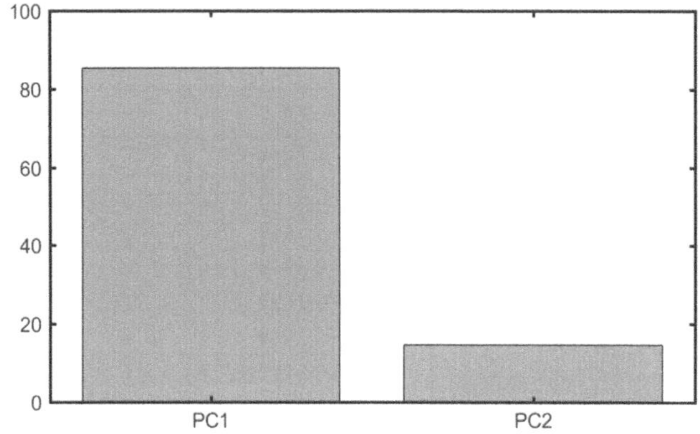


*Figure 4.3* The scree plot.

*Table 4.6* Calculation of weighted multi-response performance index (MPI)

| *Sequence no.* | *Weighted MPI* |
|---|---|
| 1 | 0.8750 |
| 2 | 0.7826 |
| 3 | 0 |
| 4 | 0.8304 |
| 5 | 0.6793 |
| 6 | 0.5794 |
| 7 | 0.7904 |
| 8 | 0.8759 |
| 9 | 0.5983 |
| 10 | 0.6157 |
| 11 | 0.7365 |
| 12 | 0.7996 |
| 13 | 0.9567 |
| 14 | 0.5063 |
| 15 | 0.4646 |
| 16 | 0.6306 |
| 17 | 1.0000 |
| 18 | 0.4274 |
| **$T_{MPI}$** | **0.6749** |

range [0 1]. The two responses are changed into a single response that is to be maximized. Table 4.6 shows the normalized values of weighted MPI.

Table 4.7 shows the response table for MPI, which show that feed has the maximum effect ($\Delta$ = 0.3049) followed by DOC ($\Delta$ = 0.2975) and speed ($\Delta$ = 0.2161). The optimal parameters are set at G2 (tool nose radius at 0.8 mm),

*Table 4.7* Response table for MPI

| *Levels* | *G* | *H* | *I* | *J* | *K* | *L* |
|---|---|---|---|---|---|---|
| L1 | 0.6679 | 0.6349 | 0.7831 | 0.7508 | 0.6525 | 0.8478 |
| L2 | 0.6819 | 0.6694 | 0.7634 | 0.7393 | 0.7306 | 0.6268 |
| L3 | Nil | 0.7204 | 0.4782 | 0.5347 | 0.6417 | 0.5502 |
| Delta | 0.0140 | 0.0855 | 0.3049 | 0.2161 | 0.0889 | 0.2975 |
| Rank | six | five | one | three | four | two |

H3 (tool rake angle at +6 degrees), I1 (feed at 0.05 mm/rev), J1 (speed at 55.42 m/min), K2 (cutting condition is wet cooling) and L1 (DOC at 0.2 mm).

The main effect of the different parameters when they are altered from the lower to higher levels can be visualized from Figure 4.4, which shows the response graph of the multiple performance index (MPI) for G, H, I, J, K, and L. It is clear from Figure 4.4 that the multiple performance index (MPI) is highest at G2, H3, I1, J1, K2, and L1. The results of ANOVA for the MPI are recorded in Table 4.8. From the ANOVA result, it is obtained that I, J, and L have major effects on quality loss. G, H, and K have no effect at a 95% confidence level. It shows that the three parameters of feed, speed, and DOC are established as the main factors for the chosen multiple performance characteristics, for the reason that their related P ratio is less than 0.05. It can be perceived from Table 4.8 that feed (I) is the most important machining parameter for affecting the multi-response due to its maximum percentage involvement (34.220%). The percentage error involvement is 23.73%, as shown in Table 4.8. The R-Sq value of 91.61% and R-Sq (adj) value of 76.24% show the efficiency of carrying out an experiment. The reasons for the result of the nearby study are explained as follows: The cutting environments (wet and cool) were very effective in

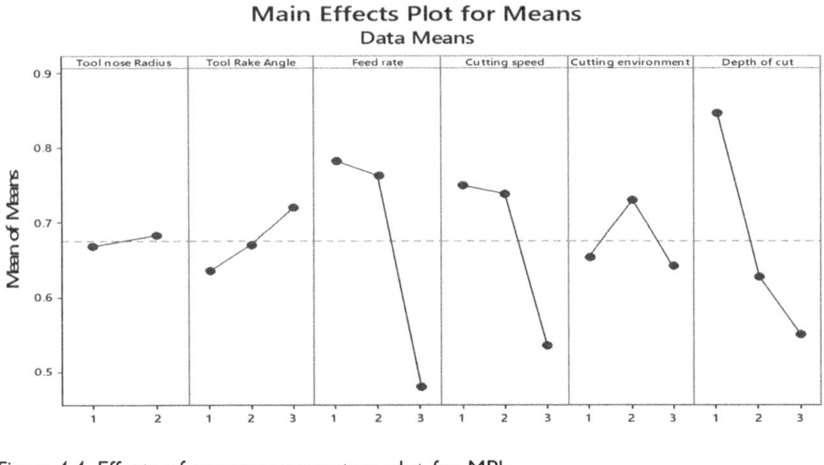


*Figure 4.4* Effects of process parameters plot for MPIs.

*Table 4.8* Analysis of variance (ANOVA) for MPI

| *Input sources* | *SS* | *DOF* | *Variance* | *F ratio* | *Prob.* | *Pure sum of square SS'* | *Percent contribution P (%)* |
|---|---|---|---|---|---|---|---|
| Nose radius of tool (G) | 0.00088 | 1 | 0.00088 | — | 0.804 | — | — |
| Rake angle of tool (H) | 0.02222 | 2 | 0.01111 | — | 0.476 | — | — |
| Feed (I) | 0.34942 | 2 | 0.17471 | 13.24* | 0.006 | 0.323 | 34.220 |
| Speed (J) | 0.17750 | 2 | 0.08875 | 6.73* | 0.029 | 0.151 | 15.997 |
| Cutting environment (K) | 0.02826 | 2 | 0.01413 | — | 0.400 | — | — |
| DOC (L) | 0.28644 | 2 | 0.14322 | 10.86* | 0.010 | 0.260 | 27.545 |
| Total error | 0.94388 | 17 | 0.01319 | | | 0.94388 | 100.00 |
| | 0.07916 | 6 | | | | 0.224 | 23.73 |

Note

* Significant at a 95% confidence level

S = 0.114861, R-Sq = 91.61%, R-Sq(adj) = 76.24%
e = error, $F_{ratio}$ = (V/error), tabulated F-ratio at 95% confidence level

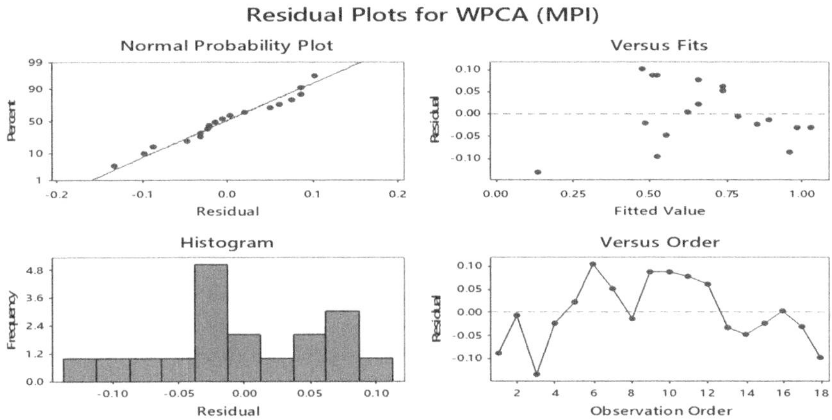


*Figure 4.5* Residual plots for WPCA (MPI).

cooling the job and tool. Figure 4.5 shows the residual plots for WPCA (MPI). The data is normal, as it lies close to the straight line n in the normal plot and other plots also show that there is no unusual structure.

## 4.5 CONFIRMATION EXPERIMENT

Just the once the most select stage of the process parameters is chosen, the last step is to calculate and confirm the performance characteristics with the supreme stage of the parameters. (Kumar et al., 2013) found I2, J2, and L1 as the optimum parameters while minimizing the surface irregularity. $\mu_{Ra}$ was 1.385 μm. The predicted confidence interval at a 95% confidence level ($CI_{CE}$) is $0.761 < \mu_{Ra}$ (microns) $< 2.009$. (Kumar et al., 2013) found I3, J3, and L3 are the optimum parameters while maximizing the metal removal rate and $\mu_{MRR}$ was 289.99 $mm^3$/sec. The predicted confidence interval at a 95% confidence level was ($CI_{CE}$) is $242.811 < \mu_{MRR}$ ($mm^3$/sec.) $< 337.169$.

Optimal design for the weighted multi-response performance index (MPI) value is established based on the expected average by considering a large amount of significant factors at their best levels. Speed, DOC, and feed are the three most important factors at the first levels, as shown in Table 4.8. In the WPCA-based Taguchi's technique, the optimal position is found to be G2H3I1J1K2L1. The predicted optimal setting becomes G2H3I1J1K2L1, as observed from Figure 4.4. The predictable mean at the optimum setting can be obtained from Equation (4.4). After assessing the optimal parameter settings, the optimal conditions are confirmed by conducting a confirmatory test. A confidence interval (CI) for the predicted mean on a verification run is calculated with Equation (4.5) (Ross, 1988):

$$\mu_{MPI} = I1 + J1 + L1 - 2_* T_{MPI} \tag{4.4}$$

$$CI_{CE} = \sqrt{F_a(1, f_e)\, V_e \left[ \frac{1}{n_{eff}} + \frac{1}{R} \right]} \tag{4.5}$$

where Fα; (1, fe) = $F_{0.05}$; (1; 6) = 5.99 (tabulated)

Total degree of freedom = 11, N =18×3 = 54,

neff = N/ {1 + [total degree of freedom]} = 54 / (1 + 11) = 4.5

R = number of repetition = 3

$T_{MPI}$ = overall mean of multi-response performance index = 0.6749 (Table 4.6).

I1, J1, and L1 are the mean values of the weighted multi-response performance index (MPI) with parameters at the most favorable levels. From Figure 4.4, $\overline{I1}$ = 0.7831, $\overline{J1}$ = 0.7508, and $\overline{L1}$ = 0.8478. Hence, $\mu_{MPI}$ = 1.0319 and Ve = error variance =0.01319 (Table 4.8).

A CI for the predicted mean on a verification run is ± 0.2093 with Equation (4.5) i.e. $CI_{CE}$ = ±0.2093.

The 95% CI of the predicted optimal weighted multi-response performance index (MPI) is: [$\mu_{MPI}$ − CI] < $\mu_{MPI}$< [$\mu_{MPI}$ + CI] i.e. 0.822 < $\mu_{MPI}$ < 1.241.

Table 4.9 shows the optimistic testing outcome. Superior conformity between the predicted machining performance and actual machining performance is shown. The mean values of the responses from these experiments are found to be within the confidence interval. The outcome has been confirmed during a confirmatory test, shown as a suitable outcome in Table 4.9.

## 4.6 CONCLUSIONS

This work is focused on efficient rotary of UD-GFRP with a carbide tool with an optimization of multiple responses. From the study, the following outcomes can be concluded:

1. WPCA combines individual principal components into a single multi-response performance index (MPI) to be used for optimization.
2. From ANOVA, the important cutting parameters that affect the surface irregularity and MRR are found to be feed rate followed by DOC and speed.
3. Multi-response optimization with the MPI is performed for rotary unidirectional-GFRP and the excellent setting of feed at 0.05 mm/rev, speed at 55.42 m/min, and DOC at 0.2 mm are found.
4. Feed is the important machining parameter that affects the multi-response with percentage involvement of 34.220%.

*Table 4.9* Confirmatory experimental results

| *Performance characteristics* | *Optimal levels (Process variables)* | *Predicted optimal value (POV) of quality characteristics* | *Predicted confidence intervals (PCI) at 95% confidence level ($CI_{CE}$)* | *Experimental results* |
|---|---|---|---|---|
| Surface roughness (Kumar et al., 2013) | G2H3I2J2K2L1 | 1.385 μm | $0.761 < \mu_{Ra}$ (microns) $< 2.009$ | 1.570 μm |
| Metal Removal Rate (Kumar et al., 2013) | G2H2I3J3K3L3 | 289.99 $mm^3$/sec | $242.811 < \mu_{MRR}$ ($mm^3$/sec) $< 337.169$ | 330.15 $mm^3$/sec |
| Weighted multi-response performance Index (WMPI) | G2H3I1J1K2L1 | 1.0319 | $0.822 < \mu_{MPI} < 1.241$ | 1.000 |

* I – feed, J –speed, and L – DOC (* significant variables)

5. The standard deviation for surface roughness and MRR are 0.5930 and 92.9546 respetively.
6. The optimal plan for WMPI is to be found in the range of 0.822 to 1.241.
7. The WPCA combined with Taguchi's process is a very successful and capable method for the optimization of multiple responses.

## ACKNOWLEDGMENT

The author acknowledges the financial support given by the NIT Kurukshetra.

## REFERENCES


Ananthakumar, P. & Ramesh, M. (2013). Optimization of turning process parameters using multivariate statistical method-PCA coupled with Taguchi method. *International Journal of Scientific Engineering and Technology*, 2(4), 263–267.

Antony, J. (2000). Multi-response optimization in industrial experiments using Taguchi's quality loss function and principal component analysis. *Quality and Reliability Engineering International*, 16, 3–8.

Bagci, E. & Işık, B. (2006). Investigation of surface roughness in turning unidirectional GFRP composites by using RS methodology and ANN. *International Journal of Advanced Manufacturing Technology*, 31, 10–17

Caggiano, A. (2018). Machining of fibre reinforced plastic composite materials. *Materials*, 11, 442.

Das, M. K., Kumar, K., Barman, T. K. & Sahoo, P. (2013). Optimization of surface roughness and MRR in EDM using WPCA. *Procedia Engineering*, 446–455.

Datta, S., Nandi, G., Bandyopadhyay, A. & Pal, P. K. (2009). Application of PCA based hybrid Taguchi method for multi-criteria optimization of submerged arc weld: A case study. *International Journal of Advanced Manufacturing Technology*, 45(3–4), 276–286.

Davim, J. P. (2009). *Machining composite materials*. London: Wiley-ISTE. ISBN: 978-1-84821-103–4.

Dorigo, M., Birattari, M. & Stutzle, T. (2006). Ant colony optimization-Artificial ants as a computational intelligence technique. *IEEE Computational Intelligence Magazine*, 1(4), 28–39.

Farmer, J. D., Packard, N. H. & Perelson, A. S. (1986). The immune system, adaptation and machine learning. *Physica*, 187–204.

Gandomi, A. H., Yang, X.-S. & Alavi, A. H. (2013). Cuckoo search algorithm: A metaheuristic approach to solve structural optimization problems. *Engineering with Computers*, 29(1), 17–35.

Gupta, M. & Kumar, S. (2015). Investigation of surface roughness and MRR for turning of UD-GFRP using PCA and Taguchi method Engineering Science and Technology. *An International Journal*, 18, 70–81.

Harvey, K. & Ansell, M. P. (2000). *Improved timber connections using bonded-in GFRP rods*. Bath, UK: Department of Materials Science and Engineering, University of Bath.

Isık, B. (2008). Experimental investigations of surface roughness in orthogonal turning of unidirectional glass-fiber reinforced plastic composite. *International Journal Advance Manufacturing Technology*, 37, 42–48.

Isik, B. & Kentli, A. (2009). Multicriteria optimization of cutting parameters in turning of UD-GFRP materials considering sensitivity. *International Journal Advance Manufacturing Technology*, 44, 1144–1153.

Kayastha, C. & Gandh, J. (2013). Optimization of process parameter in turning of copper by combination of Taguchi and principal component analysis method. *International Journal of Scientific and Research Publications*, 3(6), 1–6.

Kazmierski, C., (2012). Growth opportunities in global composites industry. Lucintel.

Khandey, U., Ghosh, S. & Hariharan, K. (2017). Machining parameters optimization for satisfying the multiple objectives in machining of MMCs. *Materials and Manufacturing Processes*, 32(10), 1082–1093.

Krishnan, K. N. & Ghose, D. (2005). Detection of multiple source locations using a glowworm metaphor with applications to collective robotics. In: *Proceedings of the Swarm Intelligence Symposium*, 84–91.

Kumar, S. & Meenu (2019). Optimization of the surface roughness and material removal rate in turning of unidirectional glass fiber reinforced plastics using the fuzzy-grey relation technique. *Indian Journal of Engineering & materials sciences*, 26, 7–19.

Kumar, S., Meenu & Satsangi, P. S. (2013). Multiple-response optimization of turning machining by the Taguchi method and the utility concept using unidirectional glass fiber-reinforced plastic composite and carbide (k10) cutting tool. *Journal of Mechanical Science and Technology*, 27(9), 2829–2837.

Mirjalili, S., Mirjalili, S. M. & Lewis, A. (2014). Grey wolf optimizer. *Advances in Engineering Software*, 69, 46–61.

Mishra, S., Sarkar, U., Taraphder, S., Datta, S., Swain, D., Saikhom, R., et al. (2017). Multivariate statistical data analysis- principal component analysis (PCA). *International Journal of Livestock Research*, 7(5), 60–78.

Mohanty, S. D., Mahapatra, S. S. & Mohanty, R. C. (2019). PCA based hybrid Taguchi philosophy for optimization of multiple responses in EDM. *Sadhana*, 44(2), 1–9.

Moshat, S., Datta, S., Bandyopadhyay, A. & Pal, P. K. (2010). Optimization of CNC end milling process parameters using PCA-based Taguchi method. *International Journal of Engineering, Science and Technology*, 2(1), 92–102

Myers, R. H. & Montgomery, D. C. (1995). *Response surface methodology process and product optimization using designed experiments*. New York, USA: Wiley.

Palanikumar, K. (2008). Application of Taguchi and response surface methodologies for surface roughness in machining glass fiber reinforced plastics by PCD tooling. *International Journal of Advanced Manufacturing Technology*, 36, 19–27.

Parida, A. K., Das, R., Sahoo, A. K. & Routara, B. C. (2014). Optimization of cutting parameters for surface roughness in machining of gfrp composites with graphite/fly ash filler. *Procedia Materials Science*, 6, 1533–1538.

Pham, D. T., Ghanbarzadeh, A., Koc, E., Otri, S., Rahim, S. & Zaidi, M. (2006). The bees algorithm-a novel tool for complex optimization problems. In: *Proceedings of the 2nd Virtual International Conference on Intelligent Production Machines and Systems*, 454–45.

Rashedi, E., Nezamabadi-Pour, H. & Saryazdi, S. (2009). GSA: A gravitational search algorithm. *Information Sciences*, 179(13), 2232–2248.

Ross, P. J. (1988). *Taguchi techniques for quality engineering*. New York: McGraw-Hills Book Company.

Ross, P. J. (1996). *Taguchi techniques for quality engineering*. New York: McGraw-Hill Book Company.

Roy, R. K. (1990). *A primer on Taguchi method*. New York: Van Nostrand Reinhold.

Su, C. T. & Tong, L. I. (1997). Multi-response robust design by principal component analysis. *Total Quality Management*, 8(6), 409–416.

Suman, B. & Kumar, P. A. (2006). Survey of simulated annealing as tool for single and multi-objective optimization. *Journal of the Operational Research Society*, 57(10), 1143–1160.

Tang, K. S., Man, K. F., Kwong, S. & He, Q. (1996). Genetic algorithms and their applications. *IEEE Signal Processing Magazine*, 13(6), 22–37.

Wang, X. M. & Zhang, L. C. (2003). An experimental investigation into the orthogonal cutting of unidirectional fibre reinforced plastics. *International Journal Machine Tools Manufacturing*, 43(10), 1015–1022.

Yang, X. S. (2012). Flower pollination algorithm for global optimization. *Unconventional Computation and Natural Computation Lecture Notes in Computer Science*, 7445, 240–249.

Chapter 5

# Processes Parameters Optimization on Surface Roughness in Turning of E-Glass UD-GFRP Composites Using Flower Pollination Algorithm (FPA)

*Surinder Kumar*[1] *and Meenu*[2]

[1]Assistant Professor, Department of Mechanical Engineering, National Institute of Technology, Kurukshetra, Haryana, India
[2]Professor, Department of Mechanical Engineering, National Institute of Technology, Kurukshetra, Haryana, India

**CONTENTS**



## 5.1 INTRODUCTION

Composites are synthetically created multiphase material having an attractive mixture of the most excellent property of the constituent phases. Usually, the matrix is continuous and surrounds the other phase. Fiber glass composite in which glass fibers are contained within polymer matrix in continuous or

DOI: 10.1201/9781003258421-6

discontinuous manner. Various common fiber glass composite applications are: automotive and aquatic bodies, plastic pipes, storage vessels and industrialized floorings. A multitude of latest applications are being used or at present investigated by the automotive industry. Machining of composite materials is a challenging task because of matrix burning, rapid tool wear, fiber pull-out, fiber fuzzing, and formation of powder like chips. Wang and Zhang [1] and Davim [2] found that the machining of fiber-reinforced polymer composite is distinct from metals due to the way the matrix material behaves and varying properties of fiber, and orientation of fiber that results in high tool wear, irregular surface finish on finished product and imperfect subsurface layer with cacks. Machining of GFRP composites is a critical operation because of incidence of fiber delamination, resin pull-out etc. Davim [3] and Davim and Pedro [4]. Many methods for machining composite materials are put forward. These include conventional and non-conventional processes such as ultrasonic machining, laser cutting, and EDM. However, their disadvantages involve a heat affected zone and lower rate of production. Considering these drawbacks, conventional machining process still finds a larger acceptance for machining composite materials.

## 5.2 LITERATURE REVIEW

Srinivasan et al. [5] established the statistical form for realizing the tool wear and important variables for the machining of GFRP composite by the ANOVA. Prasanth et al. [6] considered the machinability of UD-GFRP by utilizing brazed carbide tipped end mill tool. Taguchi $L_{25}$OA was utilized for trial proposes and ANOVA was utilized to recognize the involvement of variables on performance characteristics. Isik et al. [7] used weighting technique for turning of UD-GFRP composites with cemented carbide insert. Meenu et al. [8] used Taguchi's technique to perform experiments using polycrystalline diamond cutting insert for the machining of UD-GFRP composites and ANN was used. Kumar Surinder et al. [9] presented a fuzzy logic optimization technique to make high grey relation grade during the turning of UD-GFRP workpiece by using a PCD cutting tool. The most prominent factor was depth of cut.

Sehgal Anuj Kumar [10] presented RSM and ANN technique to improve the CNC end milling of FPD Iron Grade 80-55-06 to improve the surface roughness. The least value of surface roughness was 1.28 µm. The ANN model resulted in improved accuracy in comparison with RSM. Dautenhahn [11] proposed the use of PSO for optimization problems as it is well planned and practical. Yildiz [12] addressed hybrid robust differential evolution (HRDE) algorithm to multi-pass turning operation. From the investigation of results, it was found that the HRDE is more powerful compared to other approaches. Yang et al. [13] explored flower pollination algorithm to solve two bi-objective benchmark problems. Mohamed and Ibrahim [14] utilized the flower pollination algorithm shared with chaos theory to deal with ratios optimization problems. Jabri et al. [15] addressed hybrid genetic simulated annealing algorithm (HSAGA) to multiple passes of turning. It was observed that the planned HSAGA is more powerful than other metaheuristic algorithms.

Over the previous few decades, researchers have developed a lot of techniques and methods to resolve optimization issues. Yang et al. [16] such as genetic algorithms (GAs) Holland, [17], honeybee colony's Karaboga [18], particle swarm optimization (PSO) [19], bee algorithms (BAs) Pham et al. [20], monkey search Mucherino and Seref [21], glow worm swarm Krishnanand Ghose [22], bacterial foraging Passino [23], fish-swarm algorithm XL et al. [24], cuckoo search algorithm (CS) Gandomi, et al. [25], bat-inspired algorithm (BA) Yang [26]. These algorithms have been favourably employed to large-scale optimization problems. Yang [27] compared its achievement with GA; PSO for distinct optimization problems and found that the FPA is stronger than GA and PSO. Jerome and Natas [28] bring forward flower algorithm to solve combinatorial optimization problems. Acherjee et al. [29] utilized RSM and FPA for laser welding of dissimilar plastics. The algorithm was proficient in predicting the correct trend of the parametric effects. Alyasseri et al. [30] modified, hybridized, and tuned parameters of flower pollination to confront the complicated optimization problems. The author summarized that the flower pollination algorithm is an influential and beneficial tool for solving distinct optimization problems in a wide variety of applications.

The literature survey revealed that the machining of UD-GFRP is relatively a less invested area. In this paper, target is on the examination of optimum process parameters for the E-glass UD-GFRP composite with FPA method. In this manuscript, to optimize the parameters, viz. DOC, environment, speed, nose radius, feed and rake angle in orthogonal cutting of UD-GFRP composites for the surface roughness with FPA. Mathematical model is built using simply important parameters. A regression analysis is useful to perceive the most excellent level of cutting parameters and their importance. Unimportant parameters are not of concern. For optimization of parameters in turning, flower pollination algorithm (FPA) uses model as an objective function.

## 5.3 EXPERIMENTAL CONCEPT

### 5.3.1 Fabrication of UD-GFRP Rod and Specification

The workpiece material (GFRP rod) used consisted of UD fibers that were pulled throughout a resin bathtub into the form of the rod. The benefit of GFRP comprises [Harvey, Ansell Martin, 31] easy handling, lightweight, well-suited with resin and timber, well-suited material properties, advanced resistance to rotting, convenient in an acidic atmosphere, and abetter response due to improved resin bonding. Pultrusion processed unidirectional GFRP composite rods are used. E-glass and epoxy are the fiber and resins,

*Figure 5.1* Lathe turning center.

respectively. The details of the specification of this substance are: weight of rod 2.300 kg, length 840 mm, diameter 42 mm, reinforcement unidirectional (E' glass roving), density 1.95–2.1 gm/cc, water absorption 0.07%, tensile strength 6,500 kgf/cm$^2$, compression strength 6,000 kgf/cm$^2$, thermal conductivity 0.30 kcal/h·m$^2$·°C, glass content 75±5%, and epoxy resin content 25±5%.

### 5.3.2 Turning Process

The rotary operation is carried out using a carbide tool (K10). Figure 5.1 shows the NH22 lathe machine where the machining is performed. Measurements are made three times at each setting to record the Ra value. The rod is cut down to smaller pieces for machining. The specimens are turned at speed between 420 rpm to 1,210 rpm. The details of the carbide (K10) inserts are as shown in Table 5.1. A Tokyo seimitsu instrument is used to measure surface irregularity. The setup is displayed in Figure 5.2.

*Table 5.1* Properties of carbide (K10)

| | |
|---|---|
| Solidity | 14.95 g/cm$^3$ |
| Clearance angle | 7° |
| Flintiness | 1,600 Vickers kg/mm$^2$ |
| Transverse break strength | 2200 N/mm$^2$ |
| Thermal conductance | 95 W/M$^2$k |
| Young's modulus | 630 GPa |
| Compressive strength | 6,200 N/mm$^2$ |

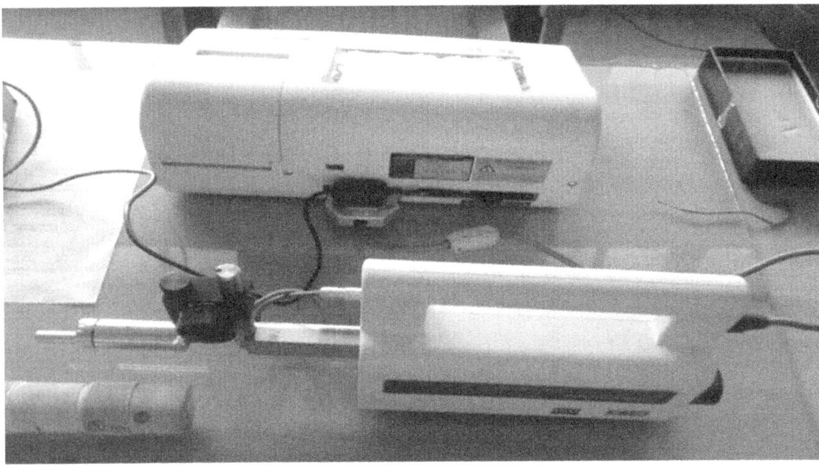

*Figure 5.2* Surface roughness tester.

## 5.4 METHODOLOGY

In this study, perfectly designed experiments by the Taguchi method are used to accumulate data. For modeling, regression methodology is used. The mathematical model is optimized using a flower pollination algorithm (FPA) method to attain the best machining situation for the necessary surface roughness.

### 5.4.1 Design of Experiments

The Taguchi methodology is a powerful technique to optimize output quality characteristics controlled by multiple process parameters Ross [32]. It assists in obtaining a combination of experiments to build experimental design easily and is well organized. Linear graphs are used to assign parameters to columns. The plan consisted of 18 tests allocated to columns 1 to 6, respectively, as shown in Table 5.2. During this study, the smaller the superior principle, it is well thought out to make the best use of the surface roughness. The signal-to-noise ratio for the response is computed [Rose, Roy, 33 and 34] as:

Smaller the best characteristics:

$$S/N = -10 \text{ Log } \frac{1}{n} \sum y^2 \tag{5.1}$$

The Taguchi $L_{18}$ orthogonal array with test conditions as specified in Table 5.3 is used. An investigational record is composed of data using accurately planned experiments. The experimental results are used to develop a second-order model.

*Table 5.2* The $L_{18}$ orthogonal array

| *Column trial* | *x1* | *x2* | *x3* | *x4* | *x5* | *x6* | — | — | *Response raw data* | | | *Signal-to-noise ratio (db)* |
|---|---|---|---|---|---|---|---|---|---|---|---|---|
| | | | | | | | | | *R1* | *R2* | *R3* | |
| 1 | 1 | 1 | 1 | 1 | 1 | 1 | 1 | 1 | $Y_{11}$ | $Y_{12}$ | $Y_{13}$ | S/N (1) |
| 2 | 1 | 1 | 2 | 2 | 2 | 2 | 2 | 2 | . | . | . | . |
| 3 | 1 | 1 | 3 | 3 | 3 | 3 | 3 | 3 | . | . | . | . |
| 4 | 1 | 2 | 1 | 1 | 2 | 2 | 3 | 3 | . | . | . | . |
| 5 | 1 | 2 | 2 | 2 | 3 | 3 | 1 | 1 | . | . | . | . |
| 6 | 1 | 2 | 3 | 3 | 1 | 1 | 2 | 2 | . | . | . | . |
| 7 | 1 | 3 | 1 | 2 | 1 | 3 | 2 | 3 | . | . | . | . |
| 8 | 1 | 3 | 2 | 3 | 2 | 1 | 3 | 1 | . | . | . | . |
| 9 | 1 | 3 | 3 | 1 | 3 | 2 | 1 | 2 | . | . | . | . |
| 10 | 2 | 1 | 1 | 3 | 3 | 2 | 2 | 1 | . | . | . | . |
| 11 | 2 | 1 | 2 | 1 | 1 | 3 | 3 | 2 | . | . | . | . |
| 12 | 2 | 1 | 3 | 2 | 2 | 1 | 1 | 3 | . | . | . | . |
| 13 | 2 | 2 | 1 | 2 | 3 | 1 | 3 | 2 | . | . | . | . |
| 14 | 2 | 2 | 2 | 3 | 1 | 2 | 1 | 3 | . | . | . | . |
| 15 | 2 | 2 | 3 | 1 | 2 | 3 | 2 | 1 | . | . | . | . |
| 16 | 2 | 3 | 1 | 3 | 2 | 3 | 1 | 2 | . | . | . | . |
| 17 | 2 | 3 | 2 | 1 | 3 | 1 | 2 | 3 | . | . | . | . |
| 18 | 2 | 3 | 3 | 2 | 1 | 2 | 3 | 1 | $Y_{181}$ | $Y_{182}$ | $Y_{183}$ | S/N (18) |

*Table 5.3* The range of process parameters

| *Parameters symbol* | *Levels* | | |
|---|---|---|---|
| | *L 1* | *L 2* | *L 3* |
| x1 | 0.4 | 0.8 | — |
| x2 | −6° | 0° | +6° |
| x3 | 0.05 | 0.1 | 0.2 |
| x4 | 55.42 | 110.84 | 159.66 |
| x5 | Dry | Wet | Cooled |
| x6 | 0.2 | 0.8 | 1.4 |

### 5.4.2 Multiple Regression Analysis

In statistics, regression studies include a lot of techniques for modeling and analyzing a number of parameters when the main concentration is on the connection between the explained variable and explanatory variables Douglas Montogomery et al. [35]. A statistical model is developed for surface irregularity. The first-order model gave low predictability i.e. high prediction error for measured responses. So, the second-order model is used for regression analysis.

### 5.4.3 Flower Pollination Algorithm (FPA)

Yang [27] developed the flower pollination algorithm (FPA), whose main objective was reproduction via pollination. Self- or cross-pollination are the two methods used for pollination. Cross-pollination also takes place at a far distance via pollinators such as bats, birds, and bees that fly a far distance and thus can be designated as the universal pollination [Pavlyukevich, 36]. Moreover, flower loyalty can be used as an addition stage using the parallel or dissimilarity of two flowers. It is considered that every plant has only single flower that only creates a single pollen gamete. The solution $x_0^d$ represents a flower in the d dimension space.

FPA has been found to be useful for economic weight dispatch problems and multi-objective optimization problems. FPA is a flexible, scalable, straightforward optimization process. Therefore, FPA reveals good quality results to solve a variety of real-life optimization efforts from dissimilar domains such as electrical and control systems, global function optimization, automatic engineering optimization, and many others Yang [27]. Characteristics of the pollination process and subsequent policies were considered for flower constancy and pollinator behavior:

1. Biotic and cross-pollination are thought of as a global pollination method with pollen-carrying pollinators carrying out Levy flights.
2. Local pollination involves abiotic and self-pollination.
3. Reproduction chance is seen as flower constancy, which is in proportion to the similarity of the two flowers taken into consideration.
4. The switch possibility controls both the local and global pollination $p$ ε [0, 1]. Local pollination can have a fraction $p$ that is considerable in the complete process of the pollination on account of physical closeness and wind speeds.

The above rules are changed into proper updating equations. For straightforwardness, it is supposed that every plant has a single flower that produces one pollen gamet. This makes sure the pollination and reproduction of the fittest solution is improved according to Rule 1 and Rule 3, as given by Equation (5.2):

$$x_i^d(t+1) = x_i^d(t) + \gamma\, L(\lambda)\, (x_i^d(t) - g_*^{d}(t)) \tag{5.2}$$

Where $x_i^d(t)$ is $i^{th}$ solution in the d dimension at iteration t and $g_*^d(t)$ is the current most suitable solution. $\gamma$ is the scaling factor to manage step size. The parameter $L(\lambda)$ is step size. A Levy flight is practiced to find the step

[Pavlyukevich, 36]. $L(\lambda) > 0$ will be drawn from a Levy distribution as insects use different steps to cover long distances, as shown in Equation (5.3):

$$L \sim \frac{\lambda T(\lambda)\sin(\pi\lambda/2)}{\pi}\frac{1}{S^{1+\lambda}},\ (S >> S_0 > 0) \tag{5.3}$$

Here, $T(\lambda)$ is the ordinary gamma function and this distribution is suitable for big steps $s > 0$; $s_0$ can be as small as 0.1. $\lambda$ is taken as 1.5. Law 2 and law 3 represent local pollination given by Equation (5.4):

$$x_i^d(t+1) = x_i^d(t) + \varepsilon(x_j^d(t) - x_k^d(t)) \tag{5.4}$$

Where $x_j^d(t)$ and $x_k^d(t)$ are result vectors drawn arbitrarily from the result set. The parameter $\varepsilon$ is drawn from even allotment in the span from 0 to 1. The nearest neighbor flowers are pollinated by local flower pollen. So, switching probability $p$ between [0 1] is used to change from global pollination to local pollination.

## 5.5 OPTIMIZATION USING THE FLOWER POLLINATION ALGORITHM AND TAGUCHI TECHNIQUE

Step-by-step procedures needed for optimizing machining variables:

1. Conversion of surface roughness data to signal-to-noise ratio. Smaller the better:

   $$\text{Signal to noise ratio} = -10\ \text{Log}\ \frac{1}{n}\sum y_{i,j}^2$$

   Where n = no. of replications
   $y_{ij}$ is the experiential response
2. Implementation of ANOVA and F-test
3. Use of regression analysis to model
4. Confirmation of model
5. Optimization with FPA. The course of action used in optimization of FPA is in Figure 5.3.

## 5.6 RESULTS AND CONVERSATION

The research trial was carried out to appraise the influence of feed, speed, nose radius, rake angle, and DOC, along with environment on the surface irregularity of the GFRP rods. Table 5.4 reveals the investigational result

**Algorithm**

Step 1: Initialize switching probability p ε [0,1]

Step 2: Generate the initial population of size N of individuals $X_i$

Step 3: Calculate the fitness $f(X_i)$ and global minimum g*

Step 4: Repeat the following steps till termination criteria is satisfied

(a) Generate the arbitrary number r between [0 1]

(b) If r<p

Update position according to Equation 2 and 3

else

Revise position according to Equation 4

(c) Evaluate new solution $X_i(t+1)$ and update the solution according to their objective value

(d) Locate the latest best answer

Step 5: Get the most excellent answer found so far

*Figure 5.3* Algorithm of FPA.

for surface irregularity that depends on the experimental parameter combinations for the 18-test condition. The least average surface irregularity of 1.570 μm is attained in experiment 17 at a nose radius of 0.8 mm, rake angle of positive 6°, feed of 0.1 mm/rev, speed of 55.42 m/min, cooled environment, and DOC of 0.2 mm. In general, a mixture of the largest level of nose radius and rake angle, moderate feed, lowest speed, and DOC and cooled environment gave rise to a better surface finish. The response table is represented in Table 5.5. Table 5.5 depicts the contribution of feed rate is highest e (Δ = 0.816) followed by cutting speed (Δ = 0.717) and DOC (Δ = 0.508).

## 5.6.1 Analysis of Variance

From Table 5.6 it is established that feed is the prime important parameter that affects surface irregularity. The insignificant parameters are rake angle, environment, and nose radius. Based on the response in Table 5.5, the selected conditions are: (1) nose radius (0.8 mm) at L2, (2) rake angle (+6 degree) at L3, (3) feed (0.1 mm/rev.) at L2, (4) speed (110.84 m/min.) at L2, (5) environment (wet) at level 2, and DOC (0.2 mm) at level 1. This analysis is executed for a level of importance of 5%. Table 5.6 reveals that the percent involvement of feed (29.110%) is enormous compared to speed (21.595%) and DOC (10.584%) in affecting the surface irregularity at a 95% assurance level.

## 5.6.2 Multiple Regression Prediction Model

A multiple regression equation is modeled for a parameter in order to calculate surface roughness for several combinations of factor levels in a particular range. So, a second-order model is used for regression analysis Equation (5.5).

*Table 5.4* Test results for surface roughness

| | Parameters | | | | | | Output values | | | | |
|---|---|---|---|---|---|---|---|---|---|---|---|
| | | | | | | | Raw data | | | Avg. $r_a$ | S/N ratio (db) |
| | | | | | | | $R_a$ (μm) | | | | |
| Exp. no. | Nose radius | Rake angle | Feed | Speed | Environment | DOC | Tr1 | Tr2 | Tr3 | | |
| 1 | 0.4 | −6° | 0.05 | 55.42 | Dry(1) | 0.2 | 1.59 | 1.65 | 1.49 | 1.577 | −3.962 |
| 2 | 0.4 | −6° | 0.1 | 110.84 | Wet(2) | 0.8 | 1.73 | 1.77 | 1.99 | 1.830 | −5.265 |
| 3 | 0.4 | −6° | 0.2 | 159.66 | Cooled (3) | 1.4 | 2.77 | 4.12 | 5.13 | 4.00 | −12.30 |
| 4 | 0.4 | 0° | 0.05 | 55.42 | 2 | 0.8 | 2.20 | 2.18 | 2.04 | 2.140 | −6.613 |
| 5 | 0.4 | 0° | 0.1 | 110.84 | 3 | 1.4 | 1.83 | 1.83 | 1.77 | 1.810 | −5.154 |
| 6 | 0.4 | 0° | 0.2 | 159.66 | 1 | 0.2 | 2.69 | 2.88 | 2.89 | 2.820 | −9.009 |
| 7 | 0.4 | +6° | 0.05 | 110.84 | 1 | 1.4 | 1.62 | 1.94 | 2.12 | 1.893 | −5.596 |
| 8 | 0.4 | +6° | 0.1 | 159.66 | 2 | 0.2 | 1.99 | 1.79 | 1.89 | 1.890 | −5.537 |
| 9 | 0.4 | +6° | 0.2 | 55.42 | 3 | 0.8 | 2.58 | 2.94 | 2.10 | 2.540 | −8.175 |
| 10 | 0.8 | −6° | 0.05 | 159.66 | 3 | 0.8 | 2.90 | 2.72 | 2.35 | 2.656 | −8.518 |
| 11 | 0.8 | −6° | 0.1 | 55.42 | 1 | 1.4 | 2.15 | 2.20 | 1.95 | 2.100 | −6.455 |
| 12 | 0.8 | −6° | 0.2 | 110.84 | 2 | 0.2 | 2.45 | 1.56 | 2.26 | 2.090 | −6.546 |
| 13 | 0.8 | 0° | 0.05 | 110.84 | 3 | 0.2 | 1.77 | 1.55 | 1.89 | 1.736 | −4.822 |
| 14 | 0.8 | 0° | 0.1 | 159.66 | 1 | 0.8 | 3.05 | 2.41 | 2.51 | 2.656 | −8.535 |
| 15 | 0.8 | 0° | 0.2 | 55.42 | 2 | 1.4 | 2.61 | 1.87 | 3.38 | 2.620 | −8.600 |
| 16 | 0.8 | +6° | 0.05 | 159.66 | 2 | 1.4 | 2.26 | 2.69 | 1.96 | 2.303 | −7.320 |
| 17 | 0.8 | +6° | 0.1 | 55.42 | 3 | 0.2 | 1.65 | 1.68 | 1.38 | 1.570 | −3.949 |
| 18 | 0.8 | +6° | 0.2 | 110.84 | 1 | 0.8 | 2.53 | 2.99 | 2.50 | 2.673 | −8.571 |
| | | | **Total** | | | | | | | **Overall Mean = 2.272** | |

*Table 5.5* Response table for surface roughness

| | *x1* | *x2* | *x3* | *x4* | *x5* | *x6* |
|---|---|---|---|---|---|---|
| L1 | 2.279 | 2.377 | 2.051 | 2.091 | 2.287 | 1.947 |
| L2 | 2.267 | 2.297 | 1.976 | 2.006 | 2.146 | 2.416 |
| L3 | — | 2.145 | 2.792 | 2.722 | 2.387 | 2.456 |
| Delta | 0.011 | 0.232 | 0.816 | 0.717 | 0.241 | 0.508 |
| Position | 6 | 5 | 1 | 2 | 4 | 3 |

*Table 5.6* ANOVA results for surface roughness

| *Input sources* | *Sum of square* | *DOF* | *Variance* | *F ratio* | *Prob.* | *Pure sum of square SS′* | *Percent contribution* |
|---|---|---|---|---|---|---|---|
| x1 | 0.0017 | 1 | 0.0017 | — | 0.922 | — | — |
| x2 | 0.4989 | 2 | 0.2495 | — | 0.245 | — | — |
| x3 | 7.3151 | 2 | 3.6575 | 21.30* | 0.000 | 6.972 | 29.110 |
| x4 | 5.5154 | 2 | 2.7577 | 16.06* | 0.000 | 5.172 | 21.595 |
| x5 | 0.5283 | 2 | 0.2641 | — | 0.227 | — | — |
| x6 | 2.8789 | 2 | 1.4394 | 8.38* | 0.001 | 2.535 | 10.584 |
| T e (pooled) | 23.9501 | 53 | | | 23.9501 | 100.00 | |
| | 7.2119 | 42 | | | 0.1717 | 9.101 | 37.99 |

Note

* Significant at a 95% confidence level.

e = error, $F_{ratio}$ = (V/error), tabulated F-ratio at 95% confidence level

$$Ra = 7.35 + 1.41\,x_1 + (-6.61)\,x_2 + 0.268\,x_3 + 0.312\,x_1x_2 + 0.133\,x_1x_3 + (-0.069)\,x_2x_3 + 0.898\,x_1^2 + 1.82\,x_2^2 + (-0.229)\,x_3^2 \quad (5.5)$$

Kumar Surinder et al. [37] $R_a$ and different predictors are taken as log values. The multiple regression coefficients $R^2$ of the second-order model are evaluated as 95.8% and relative inaccuracy involving predicted and exact values is furnished. The average absolute error percentage of 2.285 is achieved. In accordance with results, it is apparent that the highest and lowest inaccuracy percentage for surface irregularity is 9.742% and -5.263%, which is adequate. It became clear that relative inaccuracy of surface irregularity is well within reasonable bounds. Hence, this model is utilized for optimization using the flower pollination algorithm (FPA).

### 5.6.3 Optimization (FPA)

For optimization, a program is written in MATLAB. The achievement of the optimization method depends upon the arrangement of the parameters.

*Table 5.7* Optimum parameters to minimize surface roughness

| *Technique* | *Optimum parameters* | | | $R_a$ |
|---|---|---|---|---|
| | *Feed* | *Speed* | *DOC* | |
| FPA | 0.0866 | 78.2958 | 0.2000 | 1.3744 |

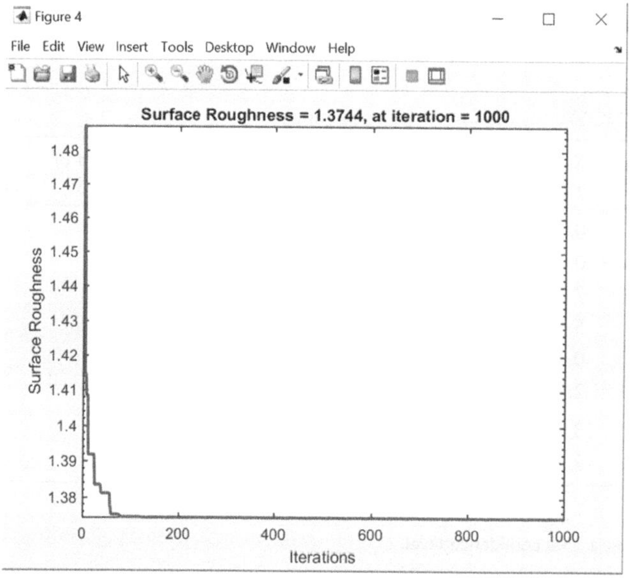


*Figure 5.4* Surface roughness verses iterations FPA.

A number of experiments are accomplished to tune these parameters. The switch probability parameter p is taken as 0.8. The population size chosen is 20. The maximum iterations are 1,000. The flower pollination algorithm is employed that mimics the impressive feature of the flower pollination. Flower constancy and levy distance pollination are the causes of escaping from local optima that produced the effectiveness of the algorithm. Flower constancy i.e. choosing same species of the flower frequently resulted in easy convergence of algorithm. Table 5.7 shows the best possible parameters for least surface irregularity and Figure 5.4 shows the surface roughness verses number of iterations for FPA. The flower pollination algorithm (FPA) technique came out with nearly the same results.

## 5.7 CONFIRMATION OF RESULTS

The reason of the verification test in this manuscript is to confirm the optimum cutting situation (x1 at L2, x2 at L3, x3 at L2, x4 at L2, x5 at

L2, and x6 at L1) that is recommended by the conducted experiment. At best possible values, significant parameters are set at an optimum level and the unimportant factors are put at a profitable level. The confirmation results are compared with the anticipated average. A confidence interval on confirmation run for the predicted mean is calculated using Equation (5.6) [Ross, 33]:

$$CI_{CE} = \sqrt{F_a(1, f_e)V_e\left[\frac{1}{n_{eff}} + \frac{1}{R}\right]} \tag{5.6}$$

From the response in Table 5.5, surface roughness value was found at the most favorable setting of x3 at level 2, x4 at level 2, and x6 at level 1 is 1.385 μm.

The predicted confidence interval at a 95% confidence level is ($CI_{CE}$) is $0.761 < \mu_{Ra}$ (microns) < 2.009. The verification trial is carried out at the best possible settings to verify the results. The parameter and their particular levels when particular response occupied into concern are shown in Table 5.8. Three verification experiments are performed at the best possible position of the machining parameters suggested by the analysis. Table 5.9 shows the confirmatory experimental results. The optimum average value of $R_a$ = 1.650 μm is attained. Table 5.4 shows that the results are strictly

*Table 5.8* Parameters and their selected levels (for optimal surface roughness)

| *Process parameters symbols* | *Optimal levels* |
|---|---|
| x1 | 0.8 mm (the irrelevant variable set at profitable levels) |
| x2 | 0 degree (set at economic levels) |
| x3 | 0.1 mm/rev |
| x4 | 110.84 m/min |
| x5 | Wet (set at profitable levels) |
| x6 | 0.8 mm |

*Table 5.9* Confirmatory experimental results

| *Performance characteristics* | *Optimal levels of process parameters symbols* | *Predicted value* | *Optimal experimental value (average of three confirmation experiments)* | *Predicted confidence Intervals at 95 % confidence level ($CI_{CE}$)* | *Experimental value* |
|---|---|---|---|---|---|
| Surface roughness | x1(L2), x2(L3), x3(L2), x4(L2), x5(L2), x6(L1) | 1.385 μm | 1.650 μm | $0.761 < \mu_{Ra}$ (microns) < 2.009 | 1.570 μm |

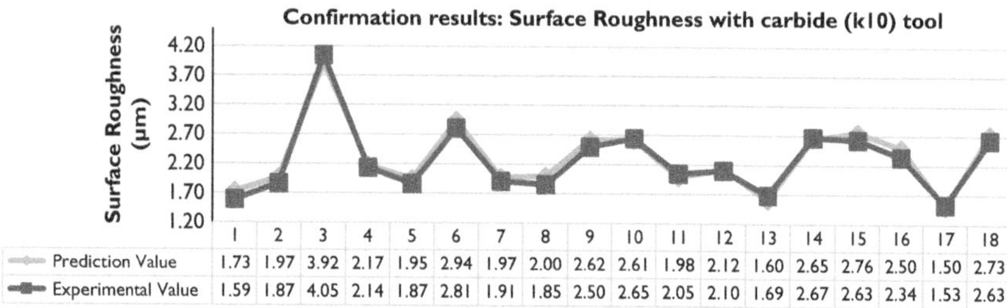


*Figure 5.5* Comparisons between real and predicted values of surface roughness.

related to the minimum average surface roughness (1.570 μm) and individual quality characteristics optimization is 1.38 μm in trial number 17. The flower pollination algorithm (FPA) technique gave about the same results as shown in Table 5.7.

Kumar Surinder et al. [38] the investigation is passed out to confirm the previously developed observed statement for surface irregularity. The two data values are closely correlated to each other signifying the validation of developed regression equation. The average absolute percentage error of 2.101 is achieved for the surface roughness model. It turns out that the highest and lowest error percentage for surface irregularity is 8.092% and −5.444%, respectively, which is very gratifying. Hence, the result predicted from the developed model is a moderately accurate graphical relationship of real and predicted values of surface irregularity and shown in Figure 5.5.

## 5.8 CONCLUSIONS

In this research, the FPA is used to improve the course parameters in the machining of E-glass UD-GFRP composites. Surface roughness is observed as an output quality characteristic. The experimentation is performed using a $L_{18}$ orthogonal array. The following summary is drawn from the present analysis:

1. In this study, single response optimization based on FPA is utilized to optimize machining variables during the machining of E-glass UD-GFRP. The experimental results indicate thatFPA is remarkably effective and has a high convergence speed and precision.
2. The ANOVA of surface roughness model shows that the predictive model is correct and believable. The main cutting parameters affecting the surface roughness are found as a feed rate followed by cutting speed. Feed is the most important factor, whilst DOC is the least important factor. The percent contribution of feed rate (29.110%) is large compared to cutting speed (21.595%) and DOC (10.584%) at a 95% assurance level).

3. Multiple regression coefficients $R^2$ value is found as 95.8%, which is very close to 1 so the developed model is highly adequate.
4. At the optimum setting of x3(L2), x4(L2), and x6(L1), the surface roughness obtained is 1.385 µm.
5. 2.101 is the average absolute percentage error obtained.
6. 1.3744 µm is the surface roughness obtained at a feed (0.0866 mm/rev), speed (78.2958 m/min) and DOC (0.2000 mm).

## ACKNOWLEDGMENTS

The authors are grateful to the NIT, Kurukshetraand MEI, India (P) Limited, Satara Maharashtra for the help offered.

## REFERENCES

1. Wang, X. M., and Zhang, L. C. An experimental investigation into the orthogonal cutting of unidirectional fibre reinforced plastics. *International Journal Machine Tools Manufacturing*, 2003, Volume 43, No. 10, pp. 1015–1022.
2. Davim, J. P. *Machining Composite Materials*. London: Wiley-ISTE, 2009.
3. Davim, P. J. A note on the determination of optimal cutting conditions on the surface finish obtained in turning using. *Journal of Material Processing Technology*, 2001, Volume 116, No. 2–3, pp. 305–308.
4. Davim, P. J., and Reis, P. Damage and dimensional precision on milling carbon fiber-reinforced plastics using design experiments. *Journal of Materials Processing Technology*, 2005, Volume 160, No. 2, pp. 160–167.
5. Srinivasan, T., Palanikumar, K., and Rajagopal, K. Roundness error evaluation in drilling of glass fiber reinforced polypropylene (GFR/PP) composites using Box Behnken Design (BBD). *Applied Mechanics and Materials*, 2015, Volume 766, No. 767, pp. 844–851.
6. Prasanth, I. S. N. V. R., Ravishankar, D. V., and Manzoor, H. M. Investigationson performance characteristics of GFRP composites in milling. *International Journal of Advanced Manufacturing Technology*, 2018, Volume 99, No. 5–8, pp. 1351–1360.
7. Isik, B., and Kentli, A. Multicriteria optimization of cutting parameters in turning of UD-GFRP materials considering sensitivity. *International Journal Advance Manufacturing Technology*, 2009, Volume 44, pp. 1144–1153.
8. Meenu, and Kumar, S. Prediction of surface roughness in turning of UD-GFRP using Artifical Neural Network. *Mechanica Confab*, 2013, Volume 2, No. 3, pp. 46–56.
9. Kumar, S., and Meenu. Optimization of the surface roughness and material removal rate in turning of unidirectional glass fiber reinforced plastics using the fuzzy-grey relation technique. *Indian Journal of Engineering & Materials Sciences*, 2019, Volume 26, pp. 7–19.
10. Kumar, S. A., and Meenu. Grey relational analysis coupled with principal component analysis to optimize the machining process of ductile iron. *Material Today*, 2018, Volume 5, No. 1, Part 1, pp. 1518–1529.

11. Dautenhahn swarm intelligence K. *Genetic Programming and Evolvable Machines*. 2002, Volume 3, No. 1, pp. 93–97.
12. Yildiz, A. R. Hybrid Taguchi-differential evolution algorithm for optimization of multi-pass turning operations. *Applied Soft Computing*, 2013, Volume 13, pp. 1433–1439.
13. Yang, X. S., and Deb, S. Cuckoo search: Recent advances and applications. *Neural Computing and Applications*, 2014, Volume 24, pp. 169–174.
14. Abdel-Baset, M., and Hezam, I. M. An improved flower pollination algorithm for ratios optimization problems. *Applied Mathematics & Information Sciences Letters*, 2015, Volume 3, No. 2, pp. 83–91.
15. Abdelouahhab, J., El Barkany, A., and El Khalfi, A. Multipass turning operation process optimization using hybrid genetic simulated annealing algorithm. *Modelling and Simulation in Engineering*, 2017, Volume 2017, pp. 1–10.
16. Yang, X. S. Flower pollination algorithm for global optimization: Unconventional Computation and Natural Computation. *Lecture Notes in Computer Science*, 2012, Volume 7445, pp. 240–249.
17. Holland, J. H. *Adaptation in Natural and Artificial Systems*. Cambridge, MA: MIT Press, 1975/1992. Second edition (1992). (First edition, University of Michigan Press, 1975).
18. Karaboga, D. An idea based on honey bee swarm for numerical optimization. *Erciyes University, Engineering Faculty, Computer Engineering Department*, 2005.
19. Kennedy, J., and Eberhart, R. Particle swarm optimization. In *Proceedings of the 1995 IEEE International Conference on Neural Networks*, December 1995, Volume 4, pp. 1942–1948.
20. Pham, D. T., Ghanbarzadeh, A., Koc, E., Otri, S., Rahim, S., and Zaidi, M. The bees algorithm-a novel tool for complex optimization problems. *Proceedings of the 2nd Virtual International Conference on Intelligent Production Machines and Systems*, 2006, pp. 454–45.
21. Mucherino, A., and Seref, O. Monkey search: A novel metaheuristic search for global optimization. In *Proceedings Data Mining, Systems Analysis and Optimization in Biomedicine*, 2007, Volume 953, pp. 162–173.
22. Krishnanand, K. N., and Ghose, D. Detection of multiple source locations using a glow-worm metaphor with applications to collective robotics. In *Proceedings of the Swarm Intelligence Symposium*, 2005, pp. 84–91.
23. Passino, K. M. Biomimicry of bacterial foraging for distributed optimization and control. *Control Systems, IEEE*, 2002, Volume 3, pp. 52–67.
24. Li, X. L., Shao, Z. J., and Qian, J. X. Optimizing method based on autonomous animats: Fish-swarm algorithm. *System Engineering Theory and Practice*, 2002, Volume 22, pp. 11–32.
25. Gandomi, A. H., Yang, X. S., and Alavi, A. H. Cuckoo search algorithm: A metaheuristic approach to solve structural optimization problems. *Engineering with Computers*, 2013, Volume 29, No. 1, pp. 17–35.
26. Yang, X. S. A new metaheuristic bat-inspired algorithm. In *Nature Inspired Cooperative Strategies for Optimization*. 2010, pp. 65–74. Berlin, Heidelberg: Springer.
27. Yang, X. S., Karamanoglu, M., and He, X. Flower pollination algorithm: A novel approach for multiobjective optimization. *Engineering Optimization*. 2014, Volume 46, No. 9, pp. 1222–1237.

28. Durand-Lose, J., and Jonoska, N. Unconventional computation and natural computation. *International Conference*. UCNC 2012 Orleans. 2012, pp. 3–7.
29. Bappa, A., Maity, D., and Kuar, A. S. Parameters optimisation of transmission laser welding of dissimilar plastics using RSM and flower pollination algorithm integrated approach. *International Journal of Mathematical Modelling and Numerical Optimisation*, 2017, Volume 8, No. 1, 1–22.
30. Alyasseri, Z. A. A., Khader, A. T., Al-Betar, M. A., Awadallah, M. A., and Yang, X.-S. Variants of the flower pollination algorithm: A review. *Study in Computational Intelligence*, 2018, pp. 91–118.
31. Harvey, K, and Ansell, M. P. *Improved Timber Connections Using Bonded-in GFRP Rods*. Bath, UK: Department of Materials Science and Engineering, University of Bath, 2000.
32. Ross, P. J. *Taguchi Method for Quality Engineering*. New York: McGraw-Hill, 1989.
33. Ross, P. J. *Taguchi Techniques for Quality Engineering*. New York: McGraw-Hills Book Company, 1988.
34. Roy, R. K. *Primer on Taguchi Method*. New York: Van Nostrand Reinhold, 1990.
35. Douglas Montogomery, C., Elizabeth, P. A., and Geoffrey Vining, G. *Introduction to Linear Regression Analysis*. AZ, USA: Arizona State University, 2001.
36. Pavlyukevich, I. Levy flights, non-local search and simulated annealing. *Journal of Computational Physics*, 2007, Volume 226, pp. 1830–1844.
37. Kumar, S., Meenu, and Satsangi, P. S. Optimization of surface roughness in turning unidirectional glass fiber reinforced plastics (UD-GFRP) composite using Carbide (K10) cutting tool. *International Journal of Advanced Design and Manufacturing Technology*, May 2013, Volume 1, No. 9, pp. 105–128.
38. Kumar, S., Meenu, Satsangi, P. S., and Sardana, H. K. Predictive modeling of surface roughness and material removal rate in turning of UD-GFRP composites using carbide (K10) tool. *International Journal of Advanced Design and Manufacturing Technology*, June 2013, Volume 6, No. 2, pp. 37–49.

Chapter 6

# Application of ANN and Taguchi Technique for Material Removal Rate by Abrasive Jet Machining with Special Abrasive Materials

*Sachin P. Ambade*[1], *Chetan K. Tembhurkar*[2], *Sagar Shelare*[2], *and Santosh Gupta*[3]

[1]Department of Mechanical Engineering, Yeshwantrao Chavan College of Engineering, Nagpur, Maharashtra, India

[2]Department of Mechanical Engineering, Priyadarshini College of Engineering, Nagpur, Maharashtra, India

[3]Department of Metallurgical and Material Engineering, Visvesvaraya National Institute of Technology (VNIT), Nagpur, Maharashtra, India

**CONTENTS**



## 6.1 INTRODUCTION

The idea of removing material from a work item with an edged cutting tool has been around for a long time. As a result, the cutting tool is moved in

DOI: 10.1201/9781003258421-7 

relation to a work piece. The material removal method is linked to plastic deformation and the creation of chips as a result. Traditionally, the cutting tool is tougher than the work material. In most circumstances, the use of such equipment and machining methods is sufficient. Science and technology are constantly evolving a novel alloys and materials with higher hardness, weight, and strengthwhich are hard to manufacture to achievea required precision and accuracy [1–7]. Many recent advances in aerospace and nuclear engineering are also related to the increasing use of diverse materials such as nitralloy, hastalloy, nimonics, waspalloy, stainless steel, carbides, metal-matrix composites, heat-resistant steels, monolithic and composite ceramics, high-performance polymers, aluminides, and so on. There is a significant demand for these materials' well-finished products with high precision and complicated intricate patterns that impact the machining process and its economics [8–10]. However, there are some hard and brittle work materials for which typical cutting procedures for material removal are insufficient. Conventional procedures, for example, make it extremely difficult to produce small holes of elaborate designs on thin brittle jobs [11–13]. Because of their low flexibility, the piercing, stamping, and extrusion processes do not perform well on brittle materials. These materials may fracture or crumble as a result of such procedures [14–16]. Drilling circular holes in brittle materials is also challenging if ordinary drills are used. For such scenarios, techniques like laser beam machining (LBM), electro-discharge machining (EDM), ultrasonic machining (USM), electron beam machining (EBM), and abrasive jet machining (AJM) are recommended [17,18]. These are known as unusual machining methods. They are unusual machining techniques because conventional tools are not used for metal cutting, there is no shear stress at the tool-workpiece contact, and some type of energy is directly utilised. Over the last 50 years, more than twenty, non-traditional production approaches have been successfully developed and used in production engineering. Table 6.1 depicts the different Non Traditional Machining (NTM) procedures [19] based on the kind of energy required for cutting.

Table 6.2 compares the process economy and their respective efficiencies, and it is obvious that AJM requires very little capital expenditure, less electricity, and is more efficient [20]. As a result, in this study, abrasive jet machining was chosen to process the most often used engineering materials.

AJM is an unconventional machining technique. The mechanical energy of compressed air and abrasive material is utilised in these operations to achieve material removal or machining. It is a material removal technology that makes use of high-speed jet of air/gas and an abrasive aggregate to remove the substance. An abrasive is a tiny, hard, irregularly shaped particle. The high-velocity jet is directed at the surface by the controllers. This machining method is based on the abrasive erosion theory. When high-velocity abrasive particles impact a hard or brittle workpiece, some metal is removed from a striking surface. This material removal method is triggered due to brittle

*Table 6.1* Non-traditional machining process classification

| *Processes* | *Mechanism ofmetalremoval* | *Energysource* | *Transfermedia* | *Typeofenergy* |
|---|---|---|---|---|
| CHM | Ablative relation | Corrosiveagent | Reactiveenvironment | Chemical |
| ECM, ECG | Iondisplacement | Highcurrent | Electrolyte | Electrochemical |
| Conventional machining | Shear | Cuttingtool | Physicalcontact | Mechanical |
| AJM, USM, WJM | Erosion | Pneumatic/HydraulicPressure | High velocityparticles | |
| EDM | Fusion | Highvoltage | Electrons | Thermoelectric |
| IBM, PAM | | Ionizedmaterial | Hotgases | |
| PAM | Vaporization | Ionizedmaterial | Ionstream | |
| LBM | | Amplifiedlight | Radiation | |

*Abrasive jet machining (AJM), Ion beam machining (IBM), Chemical machining (CHM), Laser beam machining (LBM), Electrochemical grinding (ECG), Ultrasonic machining (USM), Plasma arc machining (PAM), Electrochemical machining (ECM), Electro discharge machining (EDM), Water jet machining (WJM).

*Table 6.2* NTM approaches' process economy

| *Process* | *Investment* | *Power usage* | *Fixtures and Tooling* | *Consumption of Tool* | *Efficiency* |
|---|---|---|---|---|---|
| AJM | VeryLower | Lower | Lower | Lower | Higher |
| CHM | Average | Higher | Lower | VeryLower | Average |
| EBM | Higher | Lower | Lower | VeryLower | VeryHigher |
| ECM | VeryHigher | Average | Average | VeryLower | Lower |
| EDM | Average | Lower | Higher | Higher | Higher |
| LBM | Average | VeryLower | Lower | VeryLower | VeryHigher |
| PAM | VeryLower | VeryLower | Lower | VeryLower | VeryLower |
| USM | Lower | Lower | Lower | Average | Higher |
| Conventional machining | Lower | Lower | Lower | Lower | VeryLower |

fracture of the material (glass) as well as micro-cutting by abrasive particles. This is the basic procedure for abrasive jet machining [21–26].

AJM has been the subjected to influence of jet velocity carrier gas, abrasive size and type, stand-off distance, nozzle shape/material, and wear on total machining materials using statistical methods and numerical models with appropriate testing methodologies.

Lima et al. [27] performed AWJ studies on agate surface finish to depict the effect of process factors like abrasive mass flow rate and traverse speed on agate surface quality. The trials were carried out with four different abrasive mass flow rates and three distinct traverse speeds in two different thicknesses of agate plates, and it was discovered that the surface polish varied depending to the depth from the abrasive jet's entrance surface. Hutyrova et al. [28] demonstrated the efficacy of AJM and water jet applications of disintegrating spinning wood plastic composites (WPCs). The pressures of scientific factors (size of abrasive particles and traverse speed of cutting head) on exterior topography were studied by optical profilometry methods, and it were revealed that using the technology may successfully tackle the situation of traditional turning causing the polymer matrix melting and subsequent cutting tool adhering. Bhowmik et al. [29] created the RSM methodologyfor optimising AWJM process parameters on green composites. Confirmation tests were used to assess the model's validity and appropriateness. The suggested design may be utilised to create a methodological process for optimizing the process parameters in environmentally friendly production operations. Carach et al. [30] investigated the influence of AWJ traversal speed upon surface value as a micro-structure generated upon manufactured surface utilisingAustralian garnet as abrasive. A laser confocal microscope was utilisedfor inspection and characterisation the surfaces, and the experimental results demonstrated that coarse manufacture of complicated substances can be achieved by appropriate tool by abrasive water jet turning technology. Vasanth et al. [31] worked to understand have an impact on of operational parameters like superficial velocity, machining duration, and grain length on steel elimination rate, floor texture transformation, and floor polish in a swirling abrasive fluidized mattress machining process, a completely unique variant of Fluidized Bed Machining (FBM). Surface change changed into proven to be faster with SA-FBM than with conventional FBM, and quality abrasive grains produced a advanced floor finish.

In several industrial applications of glass fiber-reinforced plastic, Ibraheem et al. [32] employed abrasive water jet cutting technology to produce bolt holes in the construction of structural frames (GFRP). The effects of AWJM parameters upon hole forming process of woven-laminated GFRP materials were scientifically addressed to examine the consequences of the expected independent variables on the dependent variables to decide the ideal benefits of handling boundaries. The effect of dilute polymer solutions upon the roughness, width, and shape of micro-channels machined with abrasive slurry-jet micro-machining (ASJM) machinery were explored by Kowsari [33]. Even a modest quantity of a higher-molecular-weighted polymer was shown to drastically reduce the width of machined microchannels for a given jet diameter. T. Burzynski and M. Papini [34] used a ray tracing approach to account for particle size effects, mask erosive wear, feature evolution at any impact angle with second strike, curvature smoothing, and mask edge effects using the narrow band level set methodology (NB-LSM) based model. The second-strike model has been adapted to any impact angle, and it has been validated using earlier observations on covered micro-channels into glassware with ratios (AR) > 1. Same was also tested alongside earlier LSM models and computational methods. In AJM applications, Dehnadfar et al. [35] employed the shadow graphic approach to examine particle velocity and size distributions in a free jet and via a mask aperture for both angular and spherical abrasives. The influence of the mask powder form, opening size, and size upon following distribution of particle velocity and mass flux via a mask aperture was empirically proved using the shadow graphical approach. The observed mass flow patterns matched a previously published analytical model quite well, demonstrating its validity. The models were directly experimentally validated, allowing them to be employed in surface development models and to forecast a changing shape of features in microAJM.

Li et al. [36] presented and discussed a radial-mode AJM to comprehend the material removal mechanisms and process, and consequence of processing variables (abrasive mass flow rate, rotational surface speed, speed, feed, nozzle tilt angle and water pressure) on material removal rate, surface roughness and depth of cut. Whilst comparing to standard offset-mode turning, the cutting mode was more favourable for high MRR. A dimensional analysis was performed and validated and found to agree well with the experimental data, within a standard deviation of 0.2%. Fan [37] utilized molecule picture velocimetry to lead test research on molecule speeds in micro abrasive jets. Particle jet flow was discovered to have a nearly linear downstream expansion. Within range investigated, and the rate of increase likewise increased with nozzle diameter. Then, mathematical models for molecule speeds in an air jet were created, and the findings were demonstrated to match well with actual data in terms of both magnitude and variation trend H. Getu et al. [38] investigated the cryogenic and ambient temperature embedding of aluminium oxide particles in acrylonitrile butadiene styrene, polytetrafluoroethylene, poly dimethyl siloxane, and poly methyl methacrylate. The percentage area covered by inserted $Al_2O_3$ particles under cryogenic settings was found to be much lower than that of regular abrasive jet micromachining (AJM). The surface area sheltered by embedded particles was shown to decrease as embedded particle coverage increased. As a result, it was suggested that glass bead blasting at 45°C and freezing approach significantly decreased particle embedding following AJM. Various particle removal procedures were tested, and it was discovered that using cryogenic AJM reduced particle embedding.

Domiaty et al. [39] used the AJM to drill glass sheets of a range of thickness for assessing the machinability under different regulating parameters. They proposed a mathematical model and compared the results to those of other models reported before. Fast Regime-Fluidized Bed Machining (FR-FBM) was developed by Barletta et al. [40]. By altering the FR-FBM conditions, the roughness was increased by 3 to 4 times. Getu et al. [41] compared the electro-osmotic detection flow limit and separation efficiency of AJM machined-glass channels to standard wet etching with hydrofluoric acid–machined glass channels. The electro-osmotic mobility of AJM channels was found to be similar to that of wet-etched channels, and the detection limits of the two types of chips were almost identical. Surface roughness was studied in micro channels to see how it affected separation efficiency and electro-osmotic mobility. Experiment data on the impact of AJM process variables on surface roughness was also provided and analysed in order to create approaches for improving AJM surface quality. In work did by Barletta et al. [42] fly tension impacts both completing power following up on the grating and rough contact on a superficial level, it was resolved that the finished surface is profoundly dependent on it. This research also developed methods for assessing the smoothness of the inner surface of Inconel 718 tubes and enhancing their form correctness. Barletta et al. [43] developed the FB-AJM hybrid technique and conducted extensive interior polishing trials on circular tubes. With the use of a fluidized bed, a hybrid arrangement was created to stabilise the unpredictability of the abrasive flow. First, the effect of major operative factors on process was investigated using a systematic method based on design of experiments. Following that, the machining components were contemplated as far as waviness profile development and unpleasantness to progress machining productivity, consistency, and repeatability.

Some of the researchers employed AJM numerical analysis and modelling to assess the impact of jet velocity, carrier gas, stand-off distance, abrasive size and type, nozzle material, nozzle shape, and wear upon overall machining. By representing nozzle stand- off length and diverging angle of every molecule, In the review completed by Nauhi et al. [44], the disparity was created by the normal conelike dissimilarity in the stream, as well as changes in disintegration instigated by optional effects of bouncing back particles, as indicated by computational liquid elements estimations. The model expectations were very near the noticed surface examples machined in level glass and PMMA targets. Using a statistical method, Zohourkari et al. [45] evaluated the influence of important machine parameters upon the MRR in abrasive water jet turning (AWJT). The intensity of reduce and reducing head traverse velocity had been proven to be the maximum essential factors, while rotation velocity become marginal.

Kim et al. [46] created a process planning approach that generated a 3D machining route for a 3Dworkpiece and mask automatically. The application examples were also examined twice. Kim et al. [47] suggested an automated modelling approach for the 3D nonplanar mask construction based on observed geometry in micro-abrasive jet machining. To validate the algorithm, application software was created and tested using verification and actual situations. Selvan and Raju [48] cut stainless steel with AWJ. Experimental tests were directed to select the processing parameters, using regression analysis, foster an experimental model for foreseeing cut depth in abrasive water jet cutting of stainless steel, by differing grating stream rate, navigate speed, water strain and standoff distance. Using CFD and theoretical analysis, Mostofa et al. [49] enhanced component mixing in AWJ machining by blending air, water, and abrasives with in a mixing chamber. Disintegration in the spout body was more noteworthy in the main zone and expanded with spout length, as indicated by this information. Li et al. [50] laid out molecule speed models in view of molecule mean width, spout length, air thickness, wind current speed and molecule thickness in the wake of concentrating on molecule speeds in a miniature grating air fly hypothetically. By isolating the stream and the spout in air into discrete sections along the fly pivotal bearing, a mathematical answer for the models was created to register molecule speeds. The model calculations and corresponding experimental data were found to be in good agreement, with average errors of less than 4%.

Burzynski and M Papini [51] made the model of cover erosive wear in AJM that permitted them to estimate the surface improvement of both the objective and the veil by characterizing them as a ceaseless and half and half veil target surface. When wearing a mask, the glass target profiles were larger and deeper, as well as the PMMA target profiles became broader. The objective surface's advancement was determined by partitioning it into a framework of cubic cells, every one of which was appointed a harm boundary relying upon the quantity of particles that slammed into it. For a typical micro-abrasive blasting setup, the predicted eroded profiles from the simulation were compared to the empirically observed profiles, and there was excellent agreement at lower and higher particle flux [52]. S. Wan and G.C. Lim [53] developed the fine-beam cutting system by analysing the transient flow in abrasive suspension jet cutting machines after pump cutoff. The study was driven by a need to better comprehend the issue of nozzle and line clogging, which is exacerbated by use of smaller nozzles and increased operating pressures the in fine-beam systems. The form of the abrasive jet machined surface was examined by Balasubramaniam et al. [54]. The form of the surface created in AJM was determined using a semi-empirical equation. The effects of particle size, standoff distance, centre line, and peripheral velocities of the jet on the generated surface were also investigated. Verma and Lal [55] endorse a theoretical evaluation making use of a two-segment go with the drift machine to decide the abrasive particle pace in an air-abrasive jet. Numerical results were compared to experimental data and a model was developed to estimate the material removal rate in AJM (drilling) using semi-empirical fracture mechanics and a quasi-static indentation technique. The following gaps in abrasive jet machining were discovered after a thorough review of the available literature: the bulk of published research has concentrated on a certain composition of work material rather than examining numerous types of workpiece materials. Inside the fluidized bed-chamber, the uniform and homogenous mixing of abrasives with the carrier gas impinges on the whole workpiece surface over its cutting zone, which has never been documented in the literature previously.

## 6.2 EXPERIMENTATION

### 6.2.1 Development of Experimental Set-Up

Experimental set up were developed using the following components:

#### *6.2.1.1 Frame*

The machine's structure and frame are made of mild steel. The mild steel angle components were obtained and then arc welded together in the workshop to make the stand and frame. The experimental set up is depicted in Figure 6.1.

#### *6.2.1.2 Nozzle and Mixing Chamber*

The spray gun's nozzle is a component. Stainless steel is used to construct it. It was chosen because it works well with our abrasives (roasted sand, fine sand, and lime+sand). The abrasive is stored in a chamber linked to the pistol. Figure 6.2 Nozzle and Mixing Chamber.

*Figure 6.1* Frame of experimental set up.

*Figure 6.2* Nozzle and mixing chamber.

*Figure 6.3* FRL unit.

#### *6.2.1.3 FRL Unit*

An FRL device is mounted on the machine's left side to filter and control the air. It strains the compressed air, trapping solid particles (dust, filth, rust) and separating liquids (water, oil). Figure 6.3 shows FRL Unit.

#### *6.2.1.4 Funnel*

A funnel is connected underneath the workpiece to collect the used abrasives and cut particles. It has a draining effect. Figure 6.4 shows Funnel.

#### *6.2.1.5 Assembly for Movement of Nozzle*

The spray cannon is mounted on hollow rods of material to create an assembly. This permits the cannon to move in a straight line, horizontally, and vertically.

*Figure 6.4* Funnel.

#### *6.2.1.6 Mounting for Workpiece*

The mountings are used to appropriately position the workpiece and limit the workpiece's movement during experiments.

#### *6.2.1.7 Outer Cover*

To ensure operator safety, the mechanism is covered with a clear cover to prevent abrasive particles from flying out with force and injuring the operator. Figure 6.5 shows the assembly of the final developed experimental set up

### 6.2.2 Methodology for Experimentation

Soda-lime glass popular for soda–lime–silica glass, was chosen as a workpiece to test the effects of the various abrasives. The most popular variety of glass is soda-lime glass, which is used for glass containers and windowpanes for various other items. Borosilicate glass is commonly used in bakeware. Approximately 90% of all produced glass is soda–lime glass.

The standoff distance was kept constant throughout the experiments in order to achieve optimal accuracy. The three types of abrasive materials utilized for drilling were roasted sand, coarse sand, and lime + roasted sand. Experiments were carried out by altering the compressed air pressure in relation to the time necessary for drilling.

#### *6.2.2.1 Design and Parameters*

This procedure involves a large number of factors, all of which have a direct or indirect impact on the machining outcomes. For this purpose, present

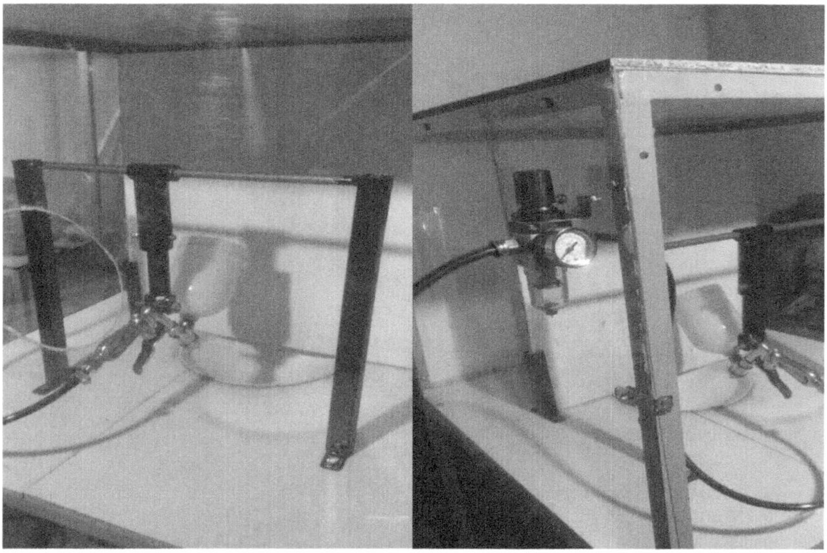

*Figure 6.5* Set-up of abrasive jet machine.

investigation only major and easy to control variables such as type of sand and pressure are considered in the experimentation. The other variables such as SOD are considered as 1 cm and kept constant during machining as shown in Tables 6.3 and 6.4. Taguchi parameter layout became utilised to attain

*Table 6.3* The variables and levels of the process

| | *Variables* | *Level 1* | *Level 2* | *Level 3* |
|---|---|---|---|---|
| A | Type of Sand | 1 | 2 | 3 |
| B | Pressure (bar) | 3.5 | 4.5 | 5.5 |

1-Roasted sand, 2-Lime + roasted sand, 3-Coarse sand

*Table 6.4* Experimental results and orthogonal array with S/N ratios

| *Sr.No.* | *Factors* | | *Time (Seconds)* | *S/N ratio* |
|---|---|---|---|---|
| | *A* | *B* | | |
| 1 | 1 | 3.5 | 15 | −23.5218 |
| 2 | 1 | 4.5 | 13 | −22.2789 |
| 3 | 1 | 5.5 | 10 | −20 |
| 4 | 2 | 3.5 | 17 | −24.4543 |
| 5 | 2 | 4.5 | 15 | −23.4052 |
| 6 | 2 | 5.5 | 13 | −22.2118 |
| 7 | 3 | 3.5 | 12 | −21.2892 |
| 8 | 3 | 4.5 | 8 | −18.2763 |
| 9 | 3 | 5.5 | 7 | −16.6502 |

excessive performance in experimental records making plans and processing. Taguchi's Signal-to-Noise (S/N) ratio became used as a statistical overall performance measure. The S/N ratio relies upon at the exceptional traits of the product/manner to be optimized. The S/N ratios is used for: - Larger-the Better, Smaller-the-better and Nominal-is-best. The parameter combination with the greatest S/N ratio is the best setting. One of the most important elements in Taguchi's approach is the choice of an orthogonal array. Orthogonal arrays are a series of all the possibilities used to determine the least number of tests. It helps to determine the optimal level of each process setting and determine the relative relevance of each process parameter. The orthogonal array L9 was selected for the experimentation. To obtain the optimum parameters combination, Taguchi's standard S /N ratios were selected (Ross, 1996) and equation as follows.

$$\eta = -10 \log_{10} [Ra^2] \quad (6.1)$$

Figure 6.6 shows the workpiece after performing the experimentation

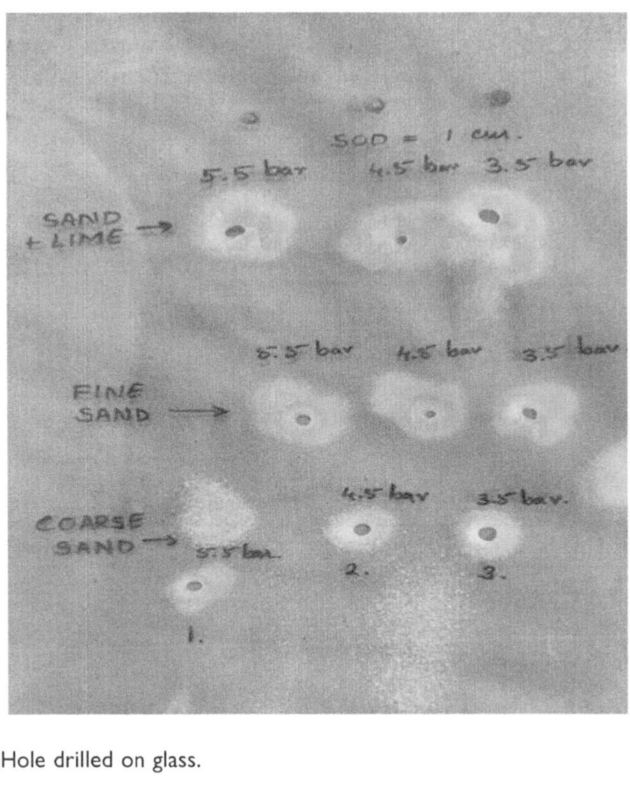


*Figure 6.6* Hole drilled on glass.

## 6.3 RESULTS & DISCUSSION

### 6.3.1 Neural Network Methodology

An artificial neural network (ANN) is designed to receive three levels of input from two elements. The elements with network diagram of the created ANN are illustrated in Figure 6.7, with the factors, kind of sand and pressure in bar are normalised on the probability scale. In a feed forward neural network, the multilayer perceptron (MLP) is chosen above convolution (CNN) and radial basis functional neural network (RNN). MLP is the simplest network, consisting of a number of simple neuron-like processing units structured in layers, with each layer's units connected to all the layers before it. The layers are joined by nodes, which form a "network" of interconnected nodes as a neural network. The network uses functions like trainlm, trainbr, trainbfg, trainrp, logsig, and trainscg to map the type of sand and pressure to the output time. These functions are used to calculate the weight (w) and bias (b). After several attempts, the size of hidden layers is set to 10. When the functions in the network achieve the minimal gradient, the training results are terminated. Table 6.5 depicts the training progress using random data division, the Levenberg-Marquardt function, and performance measurement in terms of mean squared error. Although the epoch target was set at 1000, it was reduced to 4 due to a lack of data in a short period of time. The initial mean squared error started at 0.389 and

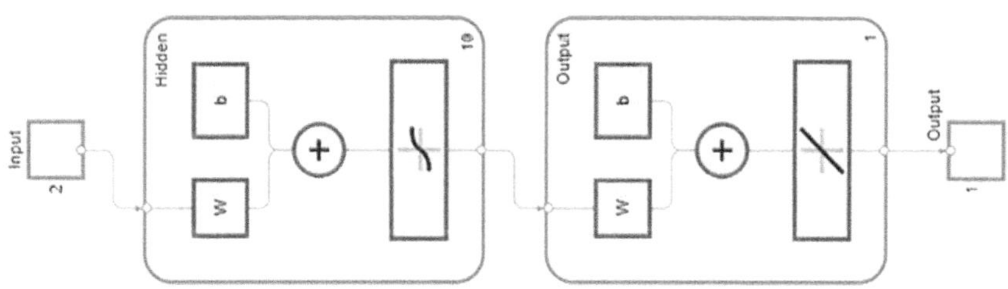


*Figure 6.7* Feed forward neural network architecture 2 × 10 × 1.

*Table 6.5* Training progress of the network with trainlm

| | *Initial* | *End* | *Target* |
|---|---|---|---|
| Epoch | 0 | 4 | 1000 |
| Elapsed Time | – | 0:00:00 | – |
| Performance | 0.389 | $5.32E^{-18}$ | 0 |
| Gradient | 0.707 | $7.45E^{-10}$ | $1.00E^{-07}$ |
| Mu | 0.001 | $1.00E^{-07}$ | $1.00E^{+10}$ |
| Validation Checks | 0 | 0 | 6 |

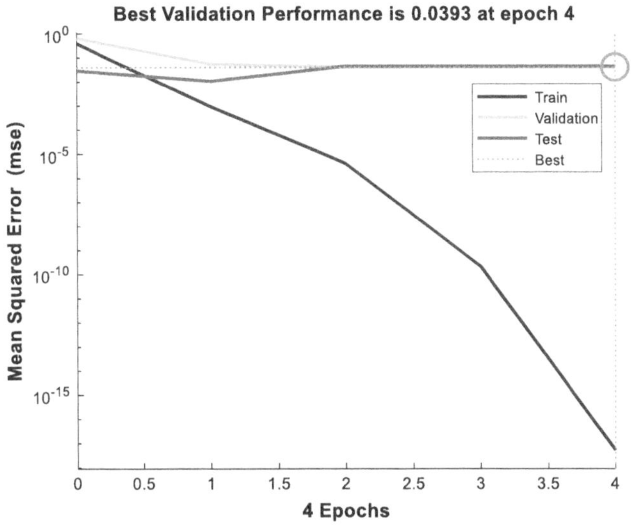


*Figure 6.8* Best validation performance with trainlm.

ended to $5.32 \times 10^{-18}$ till the convergence of the solution. The gradient as a derivative of small step function of a slope at the start was 0.707 and ended with $7.45 \times 10^{-10}$.

The momentum control parameter (mu) for the function at start was 0.001 and ended with $1 \times 10^{-7}$ to control the error convergence. The least MSE ($5.32 \times 10^{-18}$) in the validation step is happened at epoch 4 which has the best validation performance equal to 0.0393 as shown in Figure 6.8. The model training keeps going as long as the error of the network on the validation vector is reducing. The analysis stop point is at 4 epochs which is without any error repetitions as the best validation performance matches (4 epochs) with the stop point. The journey of gradient, mu and validation checks are shown in Figure 6.9. It is observed that the gradient was small at the start and increased in later epochs. However, the mu dropped linearly over the iterations. The validations checks were constant till the final epoch shows that the best validation performance (without error repetition) cannot occur before 4 epochs. The error distribution as shown in Figure 6.10 shows at seven instances during training the error were centralized with a value of 0.005461 which is almost equal to the zero error. The instances of testing and validation error were only once and on the left and right side of the training error. The training data in Figure 6.10 shows the full correlation coefficients (R values = 1). The correlation coefficient time = target value – $1.3 \times 10^{-9}$ as the best relation between the output and target values. On the other hand the validation data in Figure 6.11 failed to develop regression between output and the target values which resulted in no regression during testing. The validation and testing has R values as NaN (Not a number). The

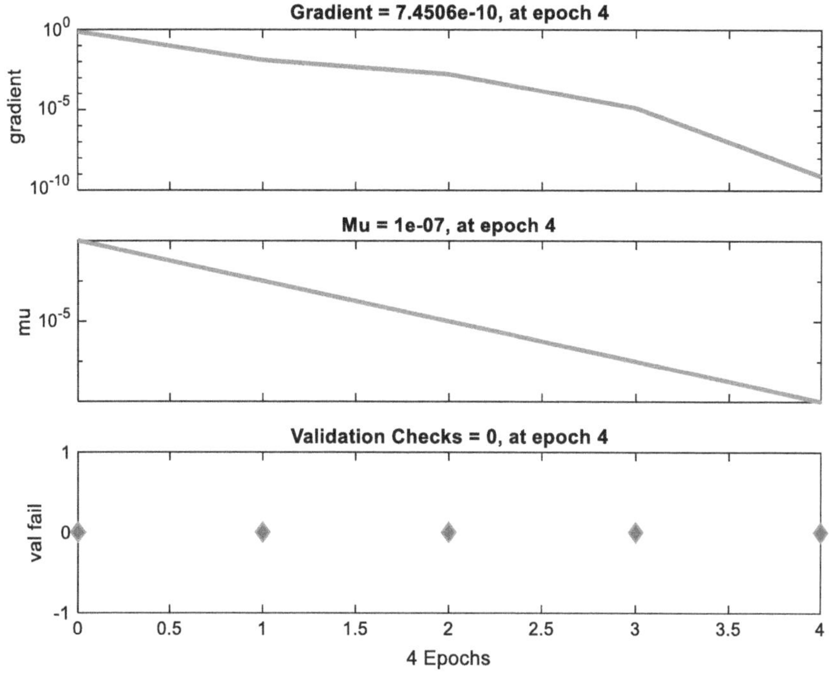


*Figure 6.9* Training state performance with trainlm.

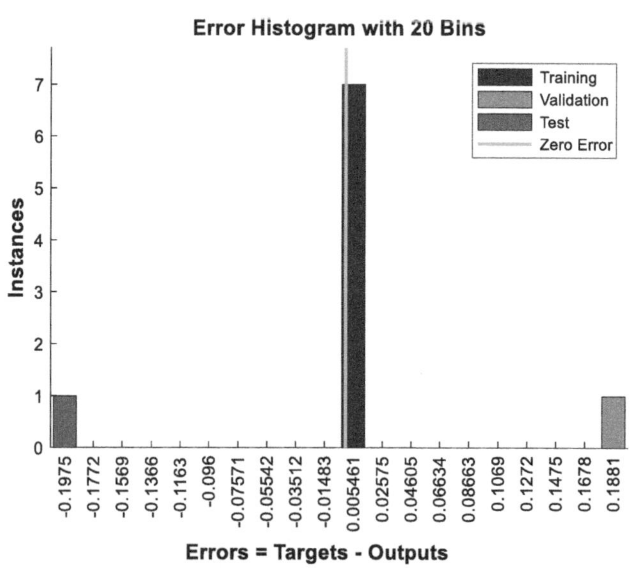


*Figure 6.10* Training error distribution with trainlm.

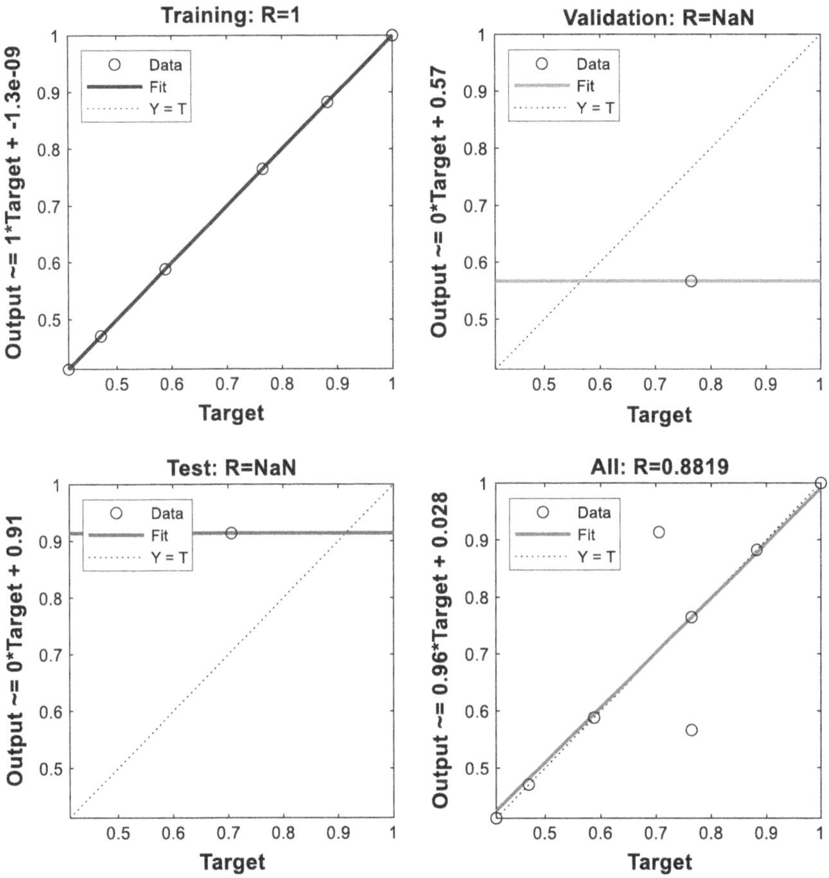


*Figure 6.11* Training regression distribution with trainbr.

overall regression thus falls down to 0.819 from full R-Values and the developed linear model for R = 0.8819 is time = 0.96 × Target values + 0.028. Figure 6.12 shows Best validation performance with trainbr, Figure 6.13 shows Training state performance with trainbr, Figure 6.14 shows Training error distribution with trainbr, Figure 6.15 shows Training regression distribution with trainbr, Figure 6.16 shows Best validation performance with trainbfg, Figure 6.17 shows Training state performance with trainbfg, Figure 6.18 shows Training error distribution with trainbfg, Figure 6.19 shows Training regression distribution with trainbfg, Figure 6.20 shows Best validation performance with trainrp, Figure 6.21 shows Training state performance with trainrp, Figure 6.22 shows Training error distribution with trainrp, Figure 6.23 shows Training regression distribution with trainrp. Tables 6.6–6.8

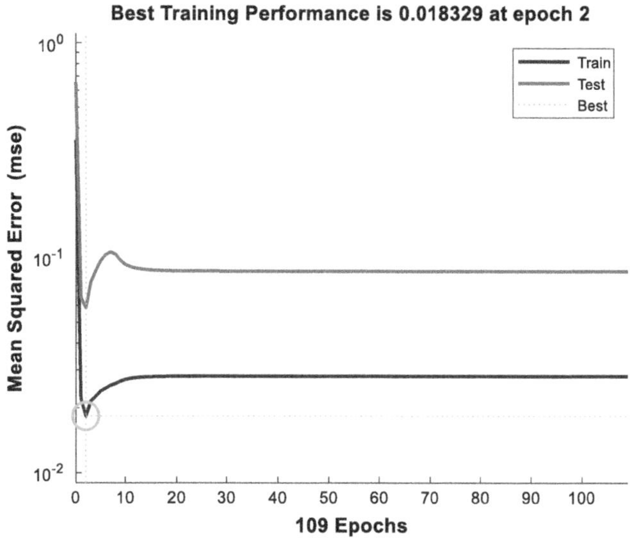


*Figure 6.12* Best validation performance with trainbr.

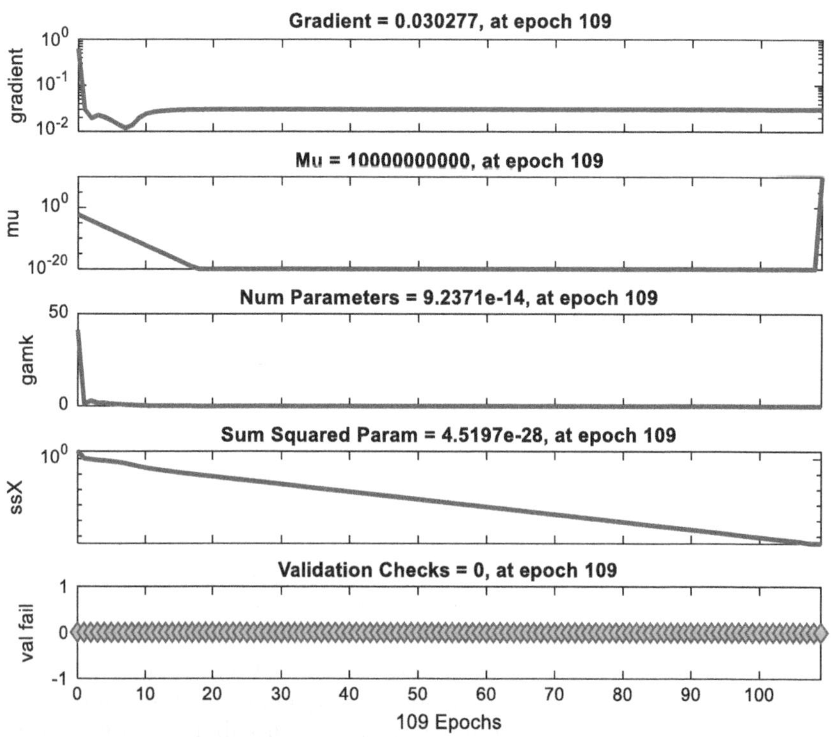


*Figure 6.13* Training state performance with trainbr.

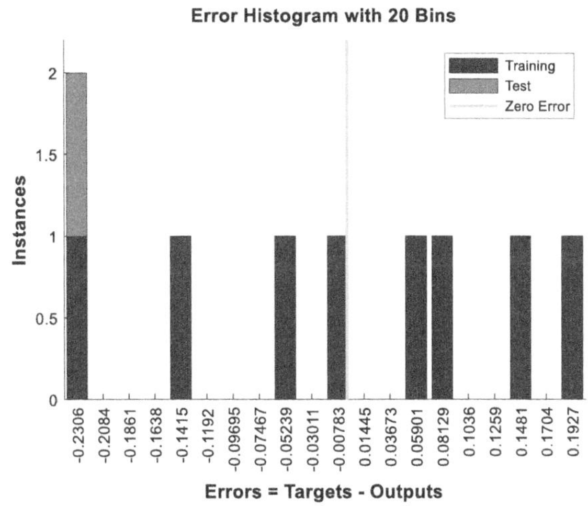


*Figure 6.14* Training error distribution with trainbr.

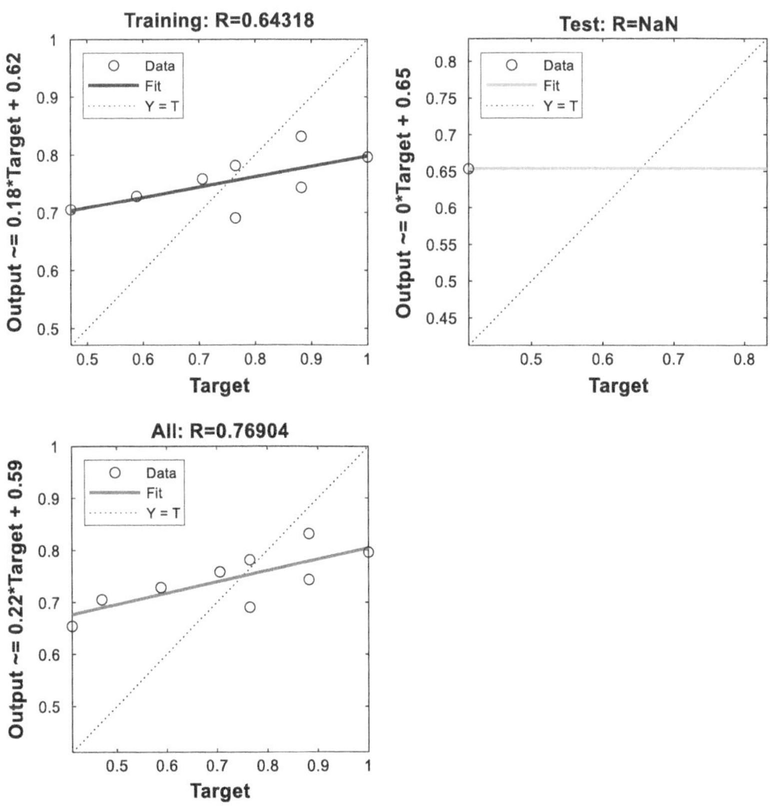


*Figure 6.15* Training regression distribution with trainbr.

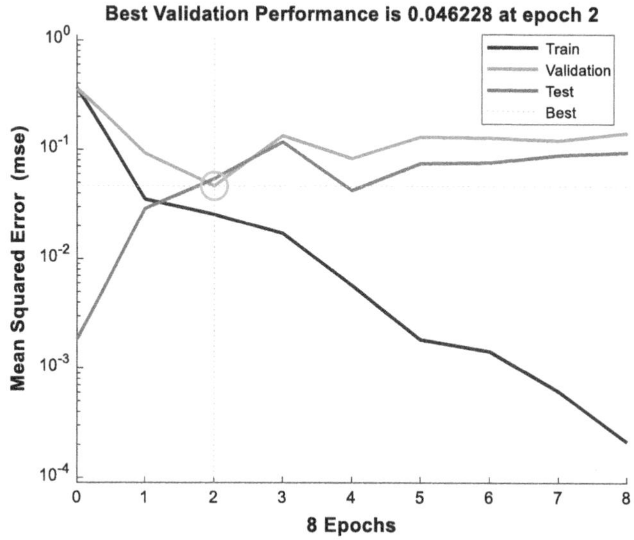


*Figure 6.16* Best validation performance with trainbfg.

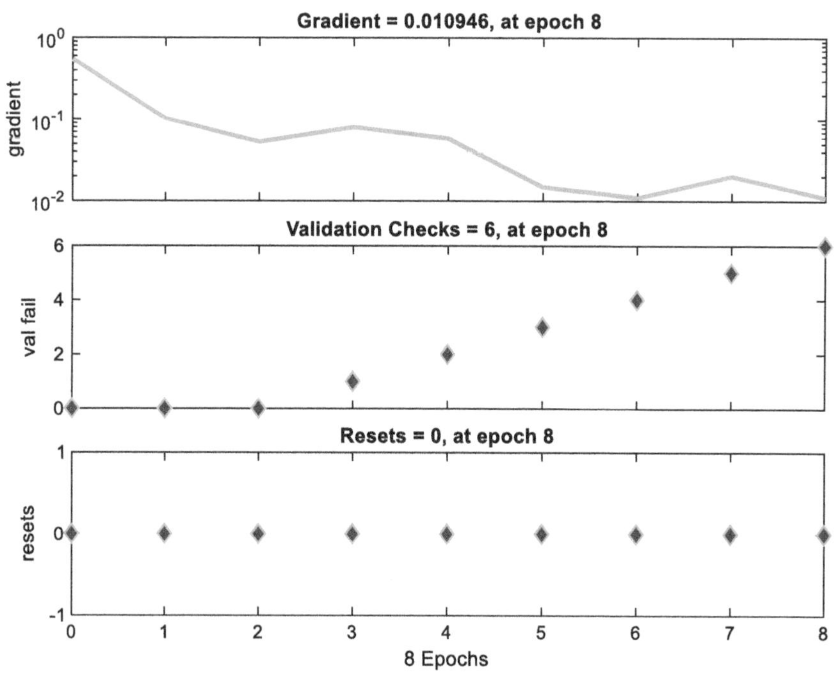


*Figure 6.17* Training state performance with trainbfg.

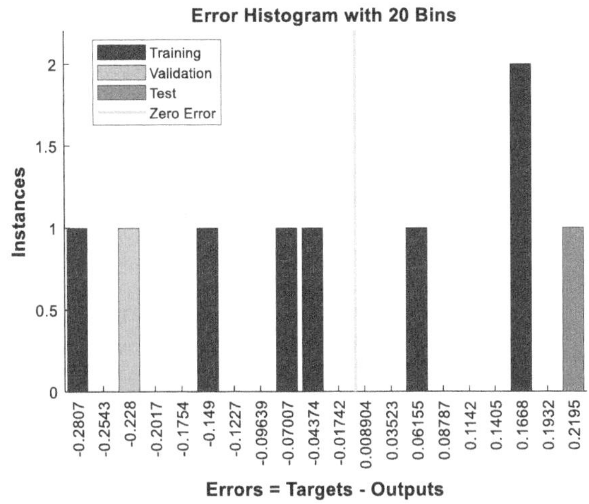


*Figure 6.18* Training error distribution with trainbfg.

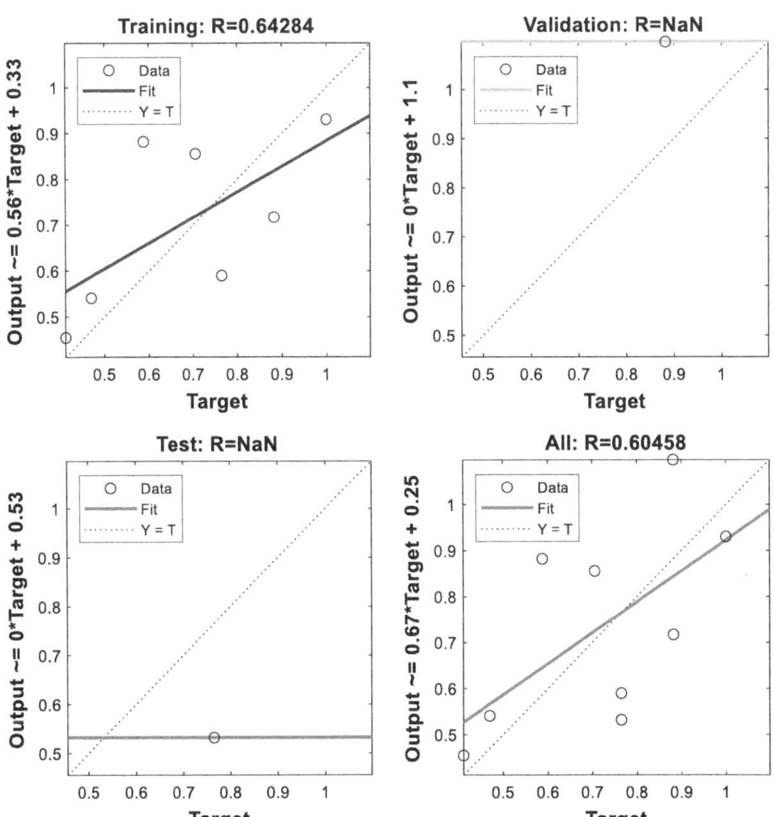


*Figure 6.19* Training regression distribution with trainbfg.

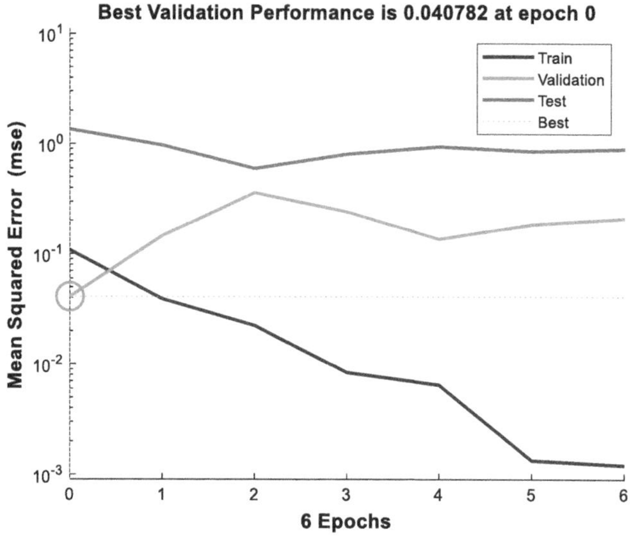


*Figure 6.20* Best validation performance with trainrp.

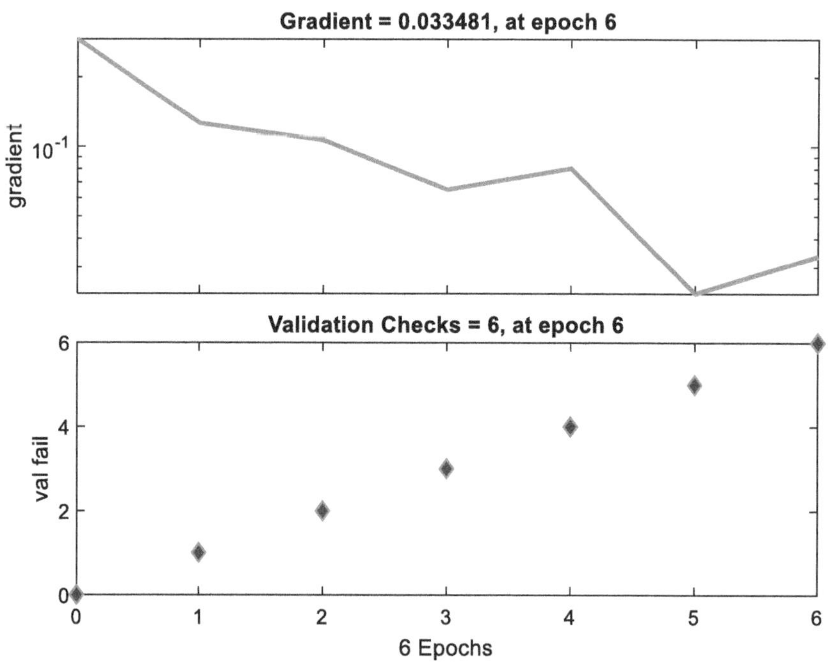


*Figure 6.21* Training state performance with trainrp.

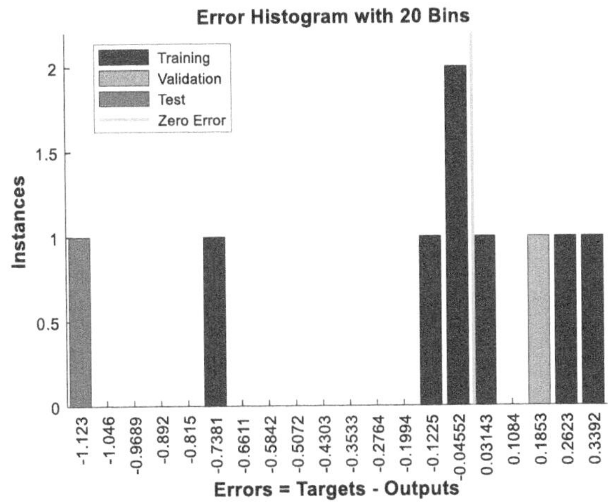


*Figure 6.22* Training error distribution with trainrp.

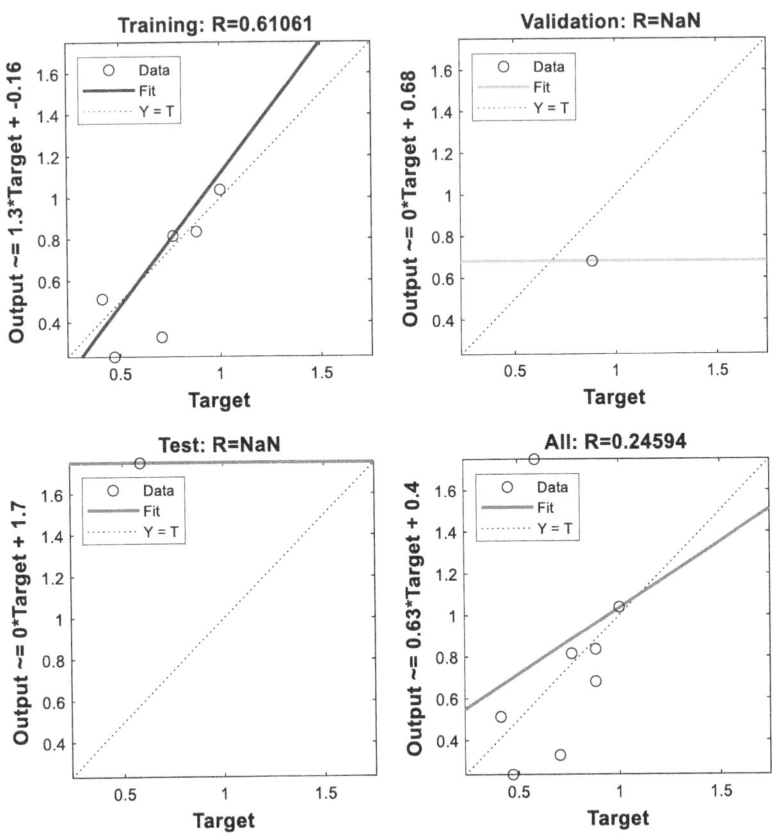


*Figure 6.23* Training regression distribution with trainrp.

*Table 6.6* Training progress of the network with trainbr

| | | | |
|---|---|---|---|
| Epoch | 0 | 109 | 1000 |
| Elapsed Time | – | 0:00:00 | – |
| Performance | 0.352 | $2.81E^{-02}$ | 0 |
| Gradient | 0.619 | $3.03E^{-02}$ | $1.00E^{-07}$ |
| Mu | 0.005 | $1.00E^{+10}$ | $1.00E^{+10}$ |
| Effective # Parameter | 41 | $9.24E^{-14}$ | 0 |
| Sum Squared Param | 281 | $4.52E^{-28}$ | 0 |

*Table 6.7* Training progress of the network with trainbfg

| | | | |
|---|---|---|---|
| Epoch | 0 | 8 | 1000 |
| Elapsed Time | – | 0:00:00 | – |
| Performance | 0.364 | $2.13E^{-04}$ | 0 |
| Gradient | 0.543 | $1.09E^{-02}$ | $1.00E^{-06}$ |
| Validation Checks | 0 | $6.00E^{+00}$ | $6.00E^{+00}$ |
| Step Size | 100 | $3.46E^{+00}$ | $1.00E^{-06}$ |
| Resets | 0 | $0.00E^{+00}$ | 4 |

*Table 6.8* Training progress of the network with trainrp

| | | | |
|---|---|---|---|
| Epoch | 0 | 6 | 1000 |
| Elapsed Time | – | 0:00:00 | – |
| Performance | 0.108 | $1.21E^{-03}$ | 0 |
| Gradient | 0.292 | $3.35E^{-02}$ | $1.00E^{-05}$ |
| Validation Checks | 0 | $6.00E^{+00}$ | $6.00E^{+00}$ |

### 6.3.2 Analysis of Single response

With the aid of the orthogonal array L9, different sands are used to test the performance of the grooving process. Equation (6.1) calculates the single-response value of S/N ratios Time (T) Table 6.9. Table 6.10 suggests the character mean values of S/N ratios of Time (T) responses. It may be determined that the most suitable combination $A_3B_3$ is biggest value of S/N ratios of Time. Therefore, $A_3B_3$ is the optimal combination of response Time.

*Table 6.9* Shows the means of the S/N ratio values across time

| *Parameter* | *Level 1* | *Level 2* | *Level 3* |
|---|---|---|---|
| Pressure | −23.085 | −21.32 | −19.62 |

*Table 6.10* Results of ANNOVA for time

| *Source* | *DF* | *Adj SS* | *Adj MS* | *F-value* | *P-value* |
|---|---|---|---|---|---|
| Pressure | 2 | 30.882 | 15.4411 | 51.85 | 0.001 |
| Type of sand | 2 | 54.196 | 27.0978 | 91.00 | 0.000 |
| Error | 4 | 1.191 | 0.2978 | | |
| Total | 8 | 86.269 | | | |

The best condition for Time is level 3 (Type 3) of sand type and level 3 (5.5 bar) of pressure, as shown in the major effect plots (Figure 6.7).

To determine which parameter has a significant influence on performance characteristics, a statistical software with an ANOVA analytical tool is used. Table 6.4 indicates the ANOVA findings for single-response S/N ratios of Time (seconds) for single-response S/N ratios of Time (seconds). Figure 6.24

It can be seen that type of sand and pressure both effects the phenomena significantly. Comparatively type of sand effects the more significantlyhaving the P value 0.000 whereas pressure have P value of 0.001. It is observed that the type of sand is one of the important factor that has more impact on machining followed by pressure on time.

As shown in Figure 6.25, the normality plot demonstrates that nearly the residuals comply with a straight line, suggesting that the recommended version is fit. The residual vs fitted plot as shown in Figure 6.26, indicates the randomized allotment on each aspect of the reference line of all residuals.

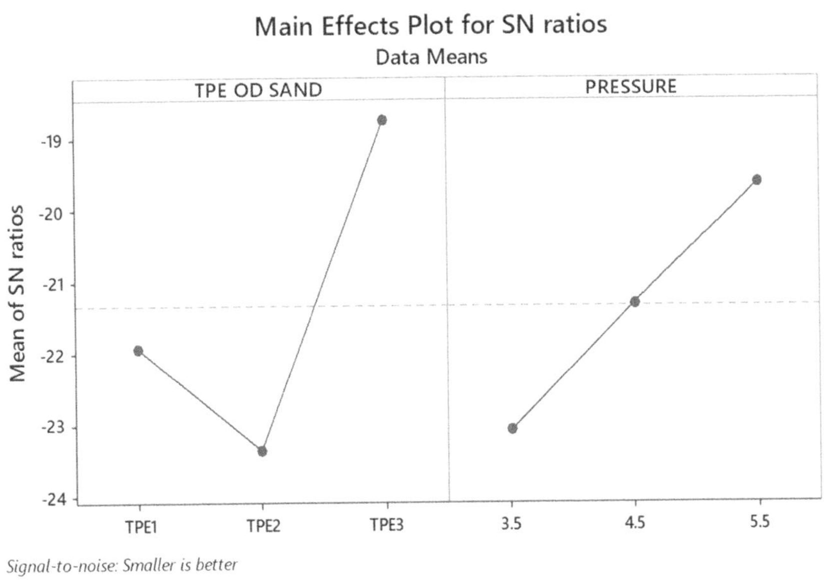


*Figure 6.24* Main effects Plots for S/N ratio of Time.

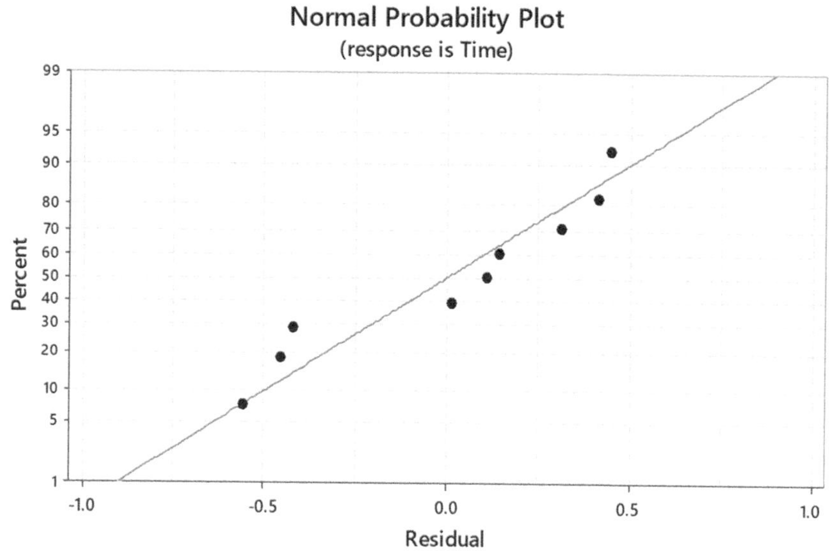


*Figure 6.25* The residual plot for a normal probability plot.

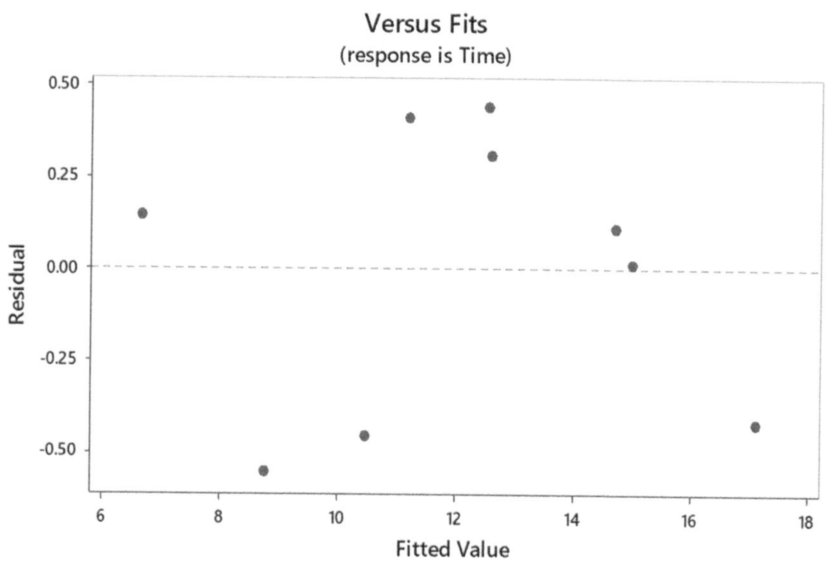


*Figure 6.26* The residual plot for a fitted vs predicted plot.

Additionally, there is no pattern in the residual vs observation order plot as per Figure 6.27, which is required for significant ANOVA. This confirms that the suggested model will have a good future outcome if all four tests are satisfied.

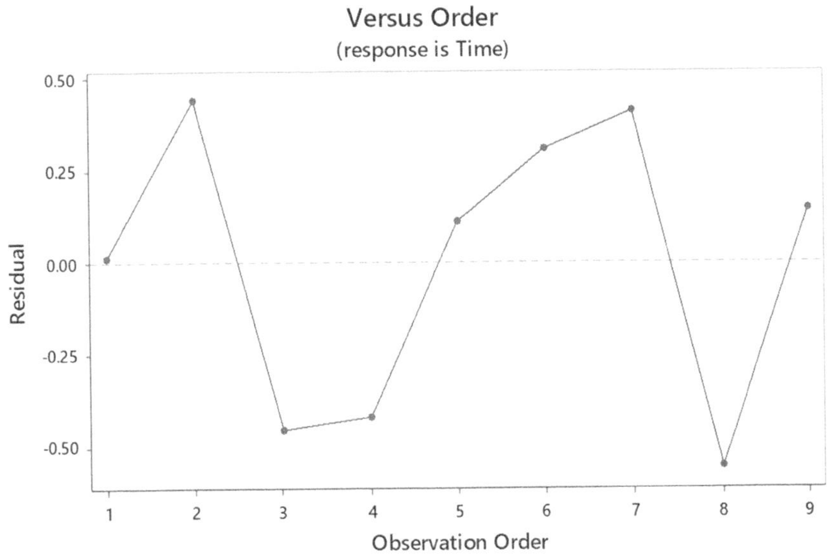


*Figure 6.27* The residual plot for a residual against observation order plot.

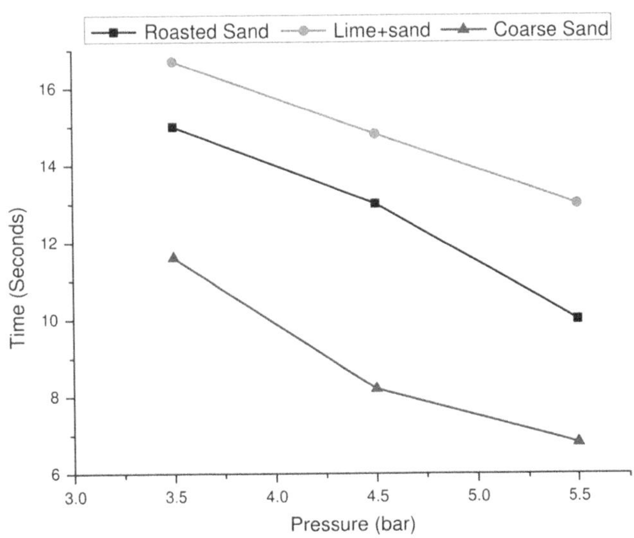


*Figure 6.28* Effect of pressure on the drilling time.

Figure 6.28 shows the effect of variation of the pressure and time required for drilling. From figure, it is clear that maximum time for the drilling is required in case of lime sand where as minimum in case of coarse sand. Maximum time of 16.7 seconds were observed at a pressure of 3.5 bar in case of lime sand whereas lowest time of 6.8 seconds at a pressure of 5.5 bar was recorded in case of coarse sand.

## 6.4 CONCLUSIONS

With increased pressure and decreasing glass thickness and SOD, the MRR rises. The pressure is proportional to the MRR. With an increase in the nozzle pressure time required for the drilling decreases continuously. By reducing the pressure from 5.5 to 3.5 bars, time increases from 10 to 15 seconds, 12.99 to 16.7, and 6.8 to 11.6 in the case of Roasted Sand, Lime+sand and Coarse Sand as an abrasive respectively. For material removal the kinetic energy of the abrasive particle is responsible during the erosion process as pressure increases. The MRR rises with an increase in SOD until a certain point, at which time it falls with an increase in SOD. As the SOD of the workpiece, nozzle jet, and abrasive mesh size reduce, the abrasive mixture impinges on the workpiece more directly, without deflecting, and with more force, the MRR increases, leading in a higher removal rate. The neural network methodology shows the overall regression thus falls down to 0.819 from full R-Values and the developed linear model for R = 0.8819 is time = 0.96 × Target values + 0.028. It shows the Best validation performance with trainbr.

## ACKNOWLEDGEMENT

The authors have grateful to the Principal Dr. U.P. Waghe, Yeshwantrao Chavan college of Engineering, Nagpur to provide the facilities for experimentation.

## REFERENCES

1. Pradhan, S., Das, S. R., & Dhupal, D. (2020). Performance evaluation of recently developed new process HAJM during machining hardstone quartz using hot silicon carbide abrasives: An experimental investigation and sustainability assessment. *Silicon*, 13(9), 2895–2919. 10.1007/s12633-020-00641-9
2. Pradhan, S., Das, S. R., Nanda, B. K., Jena, P. C., & Dhupal, D. (2020). Experimental investigation on machining of hardstone quartz with modified AJM using hot silicon carbide abrasives. *Journal of the Brazilian Society of Mechanical Sciences and Engineering*, 42(11). 1–22. 10.1007/s40430-020-02644-4
3. Chen, F., Miao, X., Tang, Y., & Yin, S. (2017). A review on recent advances in machining methods based on abrasive jet polishing (AJP). *The International Journal of Advanced Manufacturing Technology*, 90, 785–799. 10.1007/s00170-016-9405-7

4. Verma, S., Moulick, S. K., & Mishra, S. K. (2014). Nozzle wear parameter in water jet machining the review. *International Journal of Engineering Development and Research*, 2, 1063–1073. www.ijedr.org
5. Syazwani, H., Mebrahitom, G., & Azmir, A. (2016). A review on nozzle wear in abrasive water jet machining application. *IOP Conference Series: Materials Science and Engineering*, 114, 2–10. 10.1088/1757-899X/114/1/012020
6. Molitoris, M., Pitel, J., Hosovsky, A., Tothova, M., & Zidek, K. (2016). A review of research on water jet with slurry injection. *Procedia Engineering*, 149, 333–339. 10.1016/j.proeng.2016.06.675
7. Kalpana, K., Mythreyi, O. V., & Kanthababu, M. (2015). Review on condition monitoring of abrasive water jet machining system. *Proceeding of 2015 International Conference on Robotics Automation Control and Embedded Systems (RACE)*, 2015. 10.1109/RACE.2015.7097254
8. Wakuda, M., Yamauchi, Y., & Kanzaki, S. (2002). Effect of workpiece properties on machinability in abrasive jet machining of ceramic materials. *Precision Engineering*, 26, 193–198. 10.1016/S0141-6359(01)00114-3
9. Li, Z. Z., Wang, J. M., Peng, X. Q., Ho, L. T., Yin, Z. Q., Li, S. Y., & Cheung, C. F. (2011). Removal of single point diamond-turning marks by abrasive jet polishing. *Applied Optics*, 50, 2458. 10.1364/AO.50.002458
10. Haghbin, N., Spelt, J. K., & Papini, M. (2015). Abrasive waterjet micromachining of channels in metals: Comparison between machining in air and submerged in water. *International Journal of Machine Tools and Manufacture*, 88, 108–117. 10.1016/j.ijmachtools.2014.09.012
11. Li, H.-C., & Chen, W.-S. (2017). Recovery of silicon carbide from waste silicon slurry by using flotation. *Energy Procedia*, 136, 53–59. 10.1016/J.EGYPRO.2017.10.281
12. Kim, J.-Y., Kim, U.-S., Byeon, M.-S., Kang, W.-K., Hwang, K.-T., & Cho, W.-S. (2011). Recovery of cerium from glass polishing slurry. *Journal of Rare Earths*, 29, 1075–1078. 10.1016/S1002-0721(10)60601-1
13. Nouraei, H., Wodoslawsky, A., Papini, M., & Spelt, J. K. (2013). Characteristics of abrasive slurry jet micro-machining: A comparison with abrasive air jet micro-machining. *Journal of Materials Processing Technology*, 213, 1711–1724. 10.1016/j.jmatprotec.2013.03.024
14. Matsumura, T., Muramatsu, T., & Fueki, S. (2011). Abrasive water jet machining of glass with stagnation effect. *CIRP Annals – Manufacturing Technology*, 60, 355–358. 10.1016/j.cirp.2011.03.118
15. Kowsari, K., Papini, M., & Spelt, J. K. (2017). Selective removal of metallic layers from sintered ceramic and metallic plates using abrasive slurry-jet micromachining. *Journal of Manufacturing Processes*, 29, 252–264. 10.1016/j.jmapro.2017.08.005
16. Narayanan, C., Balz, R., Weiss, D. A., & Heiniger, K. C. (2013). Modelling of abrasive particle energy in water jet machining. *Journal of Materials Processing Technology*, 213, 2201–2210. 10.1016/j.jmatprotec.2013.06.020
17. Meshram, D. B., Gohil, V. K., Puri, Y. M., & Ambade, S. P. (2021). Implementation of multi-objective Jaya optimization for performance improvement in machining curve hole in P20 mold steel by sinking EDM. *World Journal of Engineering*, 19(3), 381–394. 10.1108/WJE-11-2020-0568

18. Meshram, D. B., Puri, Y. M., Gohil, V. K., & Ambade, S. P. (2020). Novel curved trajectory machining using VOEDM process – Experimental study and statistical optimization thereof. *Advances in Materials and Processing Technologies*. 10.1080/2374068X.2020.1793271
19. Su, X., Shi, L., Huang, W., & Wang, X. (2016). A multi-phase micro-abrasive jet machining technique for the surface texturing of mechanical seals. *The International Journal of Advanced Manufacturing Technology*, 86, 2047–2054.
20. Pandey, P. C., & Shan, H. S. (1980). *Modern machining processes*. Tata McGraw-Hill Education.
21. Barbatti, C., Garcia, J., Pitonak, R., Pinto, H., Kostka, A., Di Prinzio, A., Staia, M. H., & Pyzalla, A. R. (2009). Influence of micro-blasting on the microstructure and residual stresses of CVD k-Al2O3 coatings. *Surface and Coatings Technology*, 203, 3708–3717. 10.1016/j.surfcoat.2009.06.021
22. Meguid, S. A., Shagal, G., & Stranart, J. C. (1999). Finite element modelling of shotpeening residual stresses. *Journal of Materials Processing Technology*, 92–93, 401–404. 10.1016/S0924-0136(99)00153-3
23. den Dunnen, S., Dankelman, J., Kerkhoffs, G. M. M. J., & Tuijthof, G. J. M. (2016). How do jet time, pressure and bone volume fraction influence the drilling depth when waterjet drilling in porcine bone?. *Journal of the Mechanical Behavior of Biomedical Materials*, 62, 495–503. 10.1016/j.jmbbm.2016.05.030
24. Szweda, B. R. (2001). Jetting into the 21st century: Water-Jet cutting of seals and gaskets. *Sealing Technology*, 6–9.
25. Sanmartín, P., Cappitelli, F., & Mitchell, R. (2014). Current methods of graffiti removal: A review. *Construction and Building Materials*, 71, 363–374. 10.1016/j.conbuildmat.2014.08.093
26. Beaucamp, A., Namba, Y., & Freeman, R. (2012). Dynamic multiphase modeling and optimization of fluid jet polishing process. *CIRP Annals – Manufacturing Technology*, 61, 315–318. 10.1016/j.cirp.2012.03.073
27. Lima, C. E. de A., Lebrón, R., de Souza, A. J., et al. (2016). Study of influence of traverse speed and abrasive mass flow rate in abrasive water jet machining of gemstones. *The International Journal of Advanced Manufacturing Technology*, 83, 77–87.
28. Hutyrová, Z., Ščučka, J., Hloch, S., et al. (2016). Turning of wood plastic composites by water jet and abrasive water jet. *The International Journal of Advanced Manufacturing Technology*, 84, 1615–1623.
29. Bhowmik, S., & Ray, A. (2016). Prediction and optimization of process parameters of green composites in AWJM process using response surface methodology. *The International Journal of Advanced Manufacturing Technology*, 87, 1359–1370
30. Cárach, J., Hloch, S., Hlaváček, P., et al. (2016). Tangential turning of Incoloy alloy 925 using abrasive water jet technology. *The International Journal of Advanced Manufacturing Technology*, 82, 1747–1752.
31. Vasanth, S., Muthuramalingam, T., Vinothkumar, P., et al. (2016). Performance analysis of process parameters on machining titanium (Ti-6Al-4V) alloy using abrasive water jet machining process. *Procedia CIRP*, 46, 139–142.

32. Ibraheem, H. M. A., Iqbal, A., & Hashemipour, M. (2015). Numerical optimization of hole making in GFRP composite using abrasive water jet machining process. *Journal of the Chinese Institute of Engineers*, 38, 66–76.
33. Kowsari, K., Nouraei, H., James, D. F., et al. (2014). Abrasive slurry jet micromachining of holes in brittle and ductile materials. *Journal of Materials Processing Technology*, 214, 1909–1920.
34. Burzynski, T., & Papini, M. (2012). Modelling of surface evolution in abrasive jet micro-machining including particle second strikes: A level set methodology. *Journal of Materials Processing Technology*, 212, 1177–1190.
35. Dehnadfar, D., Friedman, J., & Papini, M. (2012). Laser shadowgraphy measurements of abrasive particle spatial, size and velocity distributions through micro-masks used in abrasive jet micro-machining. *Journal of Materials Processing Technology*, 212, 137–149.
36. Li, W., Zhu, H., Wang, J., et al. (2013). An investigation into the radial-mode abrasive waterjet turning process on high tensile steels. *International Journal of Mechanical Sciences*, 77, 365–376.
37. Fan, J. M., Li, H. Z., Wang, J., & Wang, C. Y. (2011). A study of the flow characteristics in micro-abrasive jets. *Experimental Thermal and Fluid Science*, 35, 1097–1106.
38. Getu, H., Spelt, J. K., & Papini, M. (2011). Reduction of particle embedding in solid particle erosion of polymers. *Wear*, 270, 922–928.
39. El-Domiaty, A., El-Hafez, H. A., & Shaker, M. A. (2009). Drilling of glass sheets by abrasive jet machining. *World Academy of Science, Engineering and Technology*, 32, 61–67.
40. Barletta, M., Rubino, G., Guarino, S., et al. (2008). Fast regime-fluidized bed machining (FR-FBM) of atmospheric plasma spraying (aps) $TiO_2$ coatings. *Surface and Coatings Technology*, 203, 855–861.
41. Getu, H., Spelt, J. K., & Papini, M. (2008). Cryogenically assisted abrasive jet micromachining of polymers. *Journal of Micromechanics and Microengineering*, 18, 115010.
42. Barletta, M., Ceccarelli, D., Guarino, S., & Tagliaferri, V. (2007). Fluidized bed assisted abrasive jet machining (FB-AJM): Precision internal finishing of Inconel 718 components. *Journal of Manufacturing Science and Engineering*, 129, 1045–1059.
43. Barletta, M., & Tagliaferri, V. (2006). Development of an abrasive jet machining system assisted by two fluidized beds for internal polishing of circular tubes. *International Journal of Machine Tools and Manufacture*, 46, 271–283.
44. Nouhi, A., Kowsari, K., Spelt, J. K., & Papini, M. (2016). Abrasive jet machining of channels on highly-curved glass and PMMA surfaces. *Wear*, 356, 30–39.
45. Zohourkari, I., Zohoor, M., & Annoni, M. (2014). Investigation of the effects of machining parameters on material removal rate in abrasive waterjet turning. *Advances in Mechanical Engineering*, 6, 624203.
46. Kim, H., Lee, I., & Ko, T. J. (2013). 3D tool path generation for micro-abrasive jet machining on 3D curved surface. *International Journal of Precision Engineering and Manufacturing*, 14, 1519–1525.

47. Kim, H., Lee, I. H., & Ko, T. J. (2012). Direct 3D mask modeling for nonplanar workpieces in microabrasive jet machining. *The International Journal of Advanced Manufacturing Technology*, 58, 175–186.
48. Selvan, M. C. P., & Raju, N. M. S. (2011). Assessment of process parameters in abrasive waterjet cutting of stainless steel. *International Journal of Advances in Engineering & Technology*, 1, 34.
49. Mostofa, M. G., Kil, K. Y., & Hwan, A. J. (2010). Computational fluid analysis of abrasive waterjet cutting head. *Journal of Mechanical Science and Technology*, 24, 249–252.
50. Li, W., Zhu, H., Wang, J., & Huang, C. (2016). Radial-mode abrasive waterjet turning of short carbon–fiber-reinforced plastics. *Machining Science and Technology*, 20, 231–248. 10.1080/10910344.2016.1165836
51. Burzynski, T., & Papini, M. (2012). A level set methodology for predicting the effect of mask wear on surface evolution of features in abrasive jet micromachining. *Journal of Micromechanics and Microengineering*, 22, 075001.
52. Shafiei, N., Getu, H., Sadeghian, A., & Papini, M. (2009). Computer simulation of developing abrasive jet machined profiles including particle interference. *Journal of Materials Processing Technology*, 209, 4366–4378.
53. Wan, S., & Lim, G. C. (2003). Transient flow in abrasive suspension jet cutting machines on pump shutdown. *Journal of Manufacturing Science and Engineering*, 125, 172–178.
54. Balasubramaniam, R., Krishnan, J., & Ramakrishnan, N. (2002). A study on the shape of the surface generated by abrasive jet machining. *Journal of Materials Processing Technology*, 121, 102–106.
55. Verma, A. P., & Lal, G. K. (1996). A theoretical study of erosion phenomenon in abrasive jet machining. *Journal of Manufacturing Science and Engineering*, 118, 564–570.

Chapter 7

# Investigation of MRR in Face Turning Unidirectional GFRP Composites by Using Multiple Regression Methodology and an Artificial Neural Network

*Surinder Kumar[1], Meenu[2], and Pawan Kumar[3]*

[1]Assistant Professor, Department of Mechanical Engineering, National Institute of Technology, Kurukshetra, Haryana, India

[2]Professor, Department of Mechanical Engineering, National Institute of Technology, Kurukshetra, Haryana, India

[3]Research Scholar, Department of Mechanical Engineering, National Institute of Technology, Kurukshetra, Haryana, India

## CONTENTS



## 7.1 INTRODUCTION

Fiber-reinforced plastic composites are two-phased materials with mechanical and thermal properties, which carry in difficult interactions among the matrix and the reinforcement throughout machining. Glass fiber–reinforced plastic composite (GFRP) brought a lot of change in the materials used in industries as they are cheap. Glass fiber–reinforced polymers are put into multiple uses that are subjected to gas, oil, and acidic environments. Machining of composites has been a difficult situation, leading to a slew of difficulties such as yarn fuzzing, matrix flaming, and shape of fine particles

DOI: 10.1201/9781003258421-8

like chips and fast tool wear, Wang and Zhang (2003) and Davim (2009). Santhanakrishnan et al. (1989) used Tic layered carbide, K20 carbide, HSS, and P20 carbide for the turning of GFRP, CFRP, and KFRP composite materials. It was discovered that the K20 carbide insert performances improved during machining of FRP composite.

Surinder Kumar et al. (2013) explored the machining of UD-GFRP composite with a carbide insert. The utility function was used for optimization. DOC, speed, and feed were remarkable factors affecting surface irregularity and MRR. Srinivas Athreya et al. (2012) used Taguchi's technique for optimization of process parameters in improving the surface irregularity of the facing process. It was detected that the cutting speed is more compelling than DOC. Karnik et al. (2008) drilled CFRP composite material by a cemented carbide (K20) drill. The ANN and trial values were compared to demonstrate the model's efficacy in predicting the delamination factor. Zain et al. (2012) considered the influence of machining parameters on the surface roughness using ANN and a genetic algorithm. Mayyas et al. (2012) found out the optimum machining parameters for drilling of self-lubricated composite material using multiple regression analysis (MRA) and ANN and the ANNs showed improved outcome compared to MRA.

Srinivas Athreya et al. (2012) applied the Taguchi process to study the effects of turning parameters on the surface roughness. It was established that the cutting speed is the most important parameter that affects surface irregularity. Meenu et al. (2013) used Taguchi's technique to perform experiments using a polycrystalline diamond cutting insert for the machining of UD-GFRP composites and ANN was used for modeling. Nizam Uddin et al. (2015) performed a face turning operation on mild steel, aluminium, and cast iron using a different amount of cutting fluid. The different materials showed a different surface finish. Dhakad et al. (2017) performed the facing of poly methyl methacrylate material using a tungsten carbide cutting tool. The experiments were performed by three-level full factorial designs by using a CNC lathe machine.

Over the last few decades, researchers have developed many techniques and methods to resolve optimization issues presented in the written text, such as particle swarm optimization (Karakuzu and Eberhart, 1995), ant colony (Dorigo et al. 2006), simulated annealing algorithm (Suman and Kumar, 2006), etc. The best possible method finds the best feasible solution for the objective function. These heuristic methods are used to discover a good-quality solution in difficult problems, as the optimal solution is not possible in practicable time because the complete area of probable solutions is to be explored (Battle, 2008). The metaheuristic algorithms use probabilistic criteria to explore the position that represents a solution and have the advantage of not being captured by local maxima. In this research effort, the machining parameters, namely speed (A), feed (B), DOC (C), and approach angle (D) are optimized for material removal rate. A comparative examination was performed using experimental values and predicted values obtained through commutative analysis and ANN. The Taguchi method under design of an experiments approach was used for experimentation and artificial neural network method under artificial intelligence was selected for modeling of parameters to maximize the MRR in turning (dry face turning) of UD-GFRP composites, which is commercially obtainable and identified for the several use in the auto industry. This method helps to acquire excellent feasible tool geometry and cutting conditions for turning of UD-GFRP with a PCD tool.

## 7.2 MATERIAL AND METHODOLOGY

### 7.2.1 Work Material

Workpiece material (GFRP rod) used consisted of unidirectional fibers that are pulled throughout a resin bathtub into the form of the rod. Because GFRP is less expensive than carbon or Kevlar, GFRP rods are employed in this work. GFRP has a number of advantages which includes (Harvey and Ansell Martin, 2000) easy handling, light weight, additionally well suited with resin and timber, high resistance to decomposition, convenient in an acidic atmosphere, and better response due to improved resin bonding. Pultrusion processed unidirectional GFRP composite rods are used. E-glass and epoxy is the fiber and resins, respectively. The properties of substance used are shown in Table 7.1. The specifications of this substance are weight of rod 2.300 kg, length 840 mm, diameter 42 mm, reinforcement unidirectional (E' Glass Roving), density 1.95–2.1 gm/cc, water absorption 0.07%, and tensile strength 6,500 kgf/cm$^2$ and compression strength 6,000 kgf/cm$^2$.

### 7.2.2 Experimental Details

The turning operation is carried out using a PCD tool of 0.8 mm tool nose radius. Figure 7.1 shows the geometry of the material of the tool. The dry facing was conducted on a NH22–lathe machine with the highest spindle velocity of 3,000 rpm and an 11 kW power motor, as shown in Figure 7.2.

*Table 7.1* Properties of UD-GFRP

| | | | |
|---|---|---|---|
| 1 | Epoxy resin content | 25±5 | % |
| 2 | Glass contented | 75±5 | // |
| 3 | Water combination | 0.07 | // |
| 4 | Strengthening, unidirectional | 'E' Glass | nill |
| 5 | Mass | 1.95–2.1 | gm/cc |
| 6 | Tensile power | 650 | (N/mm$^2$) |
| 7 | Solidity strength | 600 | // |
| 8 | Trim power | 255 | // |
| 9 | Modulus of softness | 320 | // |
| 10 | Electrical power (radial) | 3.5 | KV / mm |
| 11 | Thermal conductivity | 0.30 | Kcal /Mhc° |

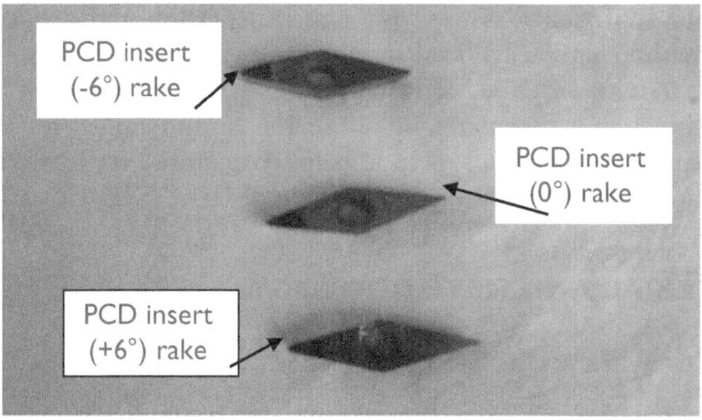


*Figure 7.1* PCD cutting tool insert.

*Figure 7.2* Experimental setup.

Four process parameters, such as cutting velocity (A), feed (B), DOC (C), and approach angle (D) are used. Measurements are made three times at each setting to record the MRR value. The electronic precision balance type instrument shown in Figure 7.3 is used for measuring the weight of MRR. The specimens after machining are shown in Figure 7.4. The MRR can be calculated as the amount of quantity of material removed in regard to

*Figure 7.3* Electronic precision balance setup.

machining time. Equations 7.1–7.2 are used for the calculation of MRR ($mm^3$/sec) and the time required for a cut is $T_c$.

$$MRR = \pi D_{avg} d \ CN \tag{7.1}$$

$$T_c = \frac{L}{CN} \tag{7.2}$$

where $D_0$ is initial and $D_i$ is the final diameter in mm

$D_{avg}$ is the avg diameter of workpiece = $\frac{(D_0 + D_i)}{2}$, $L$ is the length of cut = $\frac{D_0}{2}$, $N$ is the spindle rate in rpm, $C$ is the feed rate in mm/rev, and $d$ is the depth of cut.

### 7.2.3 Selection of Experimental Design

For modeling and analyzing the effect of variables on performance characteristics, a design of experiment is used (Arbizu and Pérez, 2003). Ranges for process parameter are specified in Table 7.2. During this study, the larger-the-better principle is considered to make the best use of the material removal rate. The S/N ratio for the response is computed as:

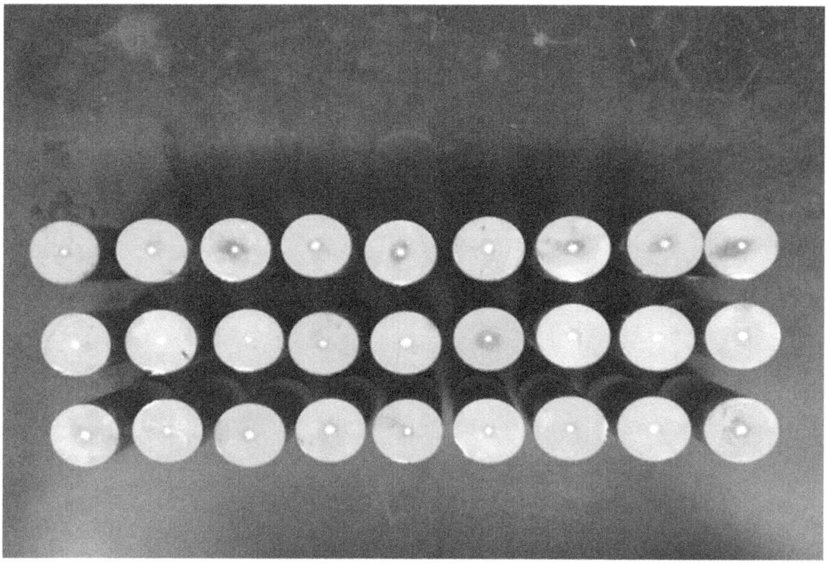

*Figure 7.4* Specimens after machining.

*Table 7.2* Control parameters and their level

| *Input variable plan* | *Levels* | | |
|---|---|---|---|
| | *L1* | *L2* | *L3* |
| Cutting velocity (rpm) A | 325 | 550 | 715 |
| Feed (mm/rev) B | 0.08 | 0.12 | 0.16 |
| DOC (mm) C | 0.6 | 0.8 | 1 |
| Approach angle (θ) D | 60 | 75 | 90 |

$$\text{Larger the better: S/N} = -10 \text{ Log } \frac{1}{n} \sum \frac{1}{y^2} \tag{7.3}$$

The Taguchi $L_{27}$ orthogonal array with test conditions as specified in Table 7.3 is used. An investigational record is composed using an accurately planned experiment. The experimental results are used to develop a second-order model.

### 7.2.4 Multiple Regression Methodology

In statistics, regression studies cover a wide range of techniques for modeling and analyzing a large number of variables, with the focus on the relationship between the explained and explanatory factors. Douglas Montogomery et al. (2001). A mathematical model is developed for material removal rate. The regression equation correlates the explained variable with the explanatory

*Table 7.3* Taguchi's $L_{27}$ orthogonal array with test conditions

| *Trial no.* | *Factor* | | | | | |
|---|---|---|---|---|---|---|
| | *Speed* | *Feed* | *DOC* | *Angle* | *MRR* | *S/N* |
| 1 | 325 | 0.08 | 0.6 | 60 | 0.03 | −33.52 |
| 2 | 325 | 0.08 | 0.8 | 75 | 0.02 | −31.45 |
| 3 | 325 | 0.08 | 1 | 90 | 0.04 | −29.74 |
| 4 | 325 | 0.12 | 0.6 | 75 | 0.05 | −32.23 |
| 5 | 325 | 0.12 | 0.8 | 90 | 0.05 | −27.12 |
| 6 | 325 | 0.12 | 1 | 60 | 0.05 | −26.56 |
| 7 | 325 | 0.16 | 0.6 | 90 | 0.06 | −25.94 |
| 8 | 325 | 0.16 | 0.8 | 60 | 0.03 | −26.53 |
| 9 | 325 | 0.16 | 1 | 75 | 0.05 | −23.96 |
| 10 | 550 | 0.08 | 0.6 | 60 | 0.06 | −29.59 |
| 11 | 550 | 0.08 | 0.8 | 75 | 0.06 | −26.68 |
| 12 | 550 | 0.08 | 1 | 90 | 0.07 | −24.99 |
| 13 | 550 | 0.12 | 0.6 | 75 | 0.09 | −24.90 |
| 14 | 550 | 0.12 | 0.8 | 90 | 0.07 | −23.27 |
| 15 | 550 | 0.12 | 1 | 60 | 0.09 | −21.26 |
| 16 | 550 | 0.16 | 0.6 | 90 | 0.11 | −23.05 |
| 17 | 550 | 0.16 | 0.8 | 60 | 0.03 | −20.87 |
| 18 | 550 | 0.16 | 1 | 75 | 0.05 | −19.33 |
| 19 | 715 | 0.08 | 0.6 | 60 | 0.05 | −30.29 |
| 20 | 715 | 0.08 | 0.8 | 75 | 0.05 | −26.21 |
| 21 | 715 | 0.08 | 1 | 90 | 0.07 | −25.86 |
| 22 | 715 | 0.12 | 0.6 | 75 | 0.08 | −25.23 |
| 23 | 715 | 0.12 | 0.8 | 90 | 0.07 | −23.20 |
| 24 | 715 | 0.12 | 1 | 60 | 0.10 | −21.90 |
| 25 | 715 | 0.16 | 0.6 | 90 | 0.11 | −23.59 |
| 26 | 715 | 0.16 | 0.8 | 60 | 0.03 | −20.16 |
| 27 | 715 | 0.16 | 1 | 75 | 0.02 | −19.20 |

variable, such as cutting speed, feed and DOC, and approach angle. The first-order linear model equation used is represented as follows:

$$\hat{Y} = b_0 + b_1x_1 + b_2x_2 + b_3x_3 + b_4x_4$$

### 7.2.5 Artificial Neural Network

ANN [2009] mimics the human brain and is a kind of artificial intelligence technique. ANNs are information processing systems and consist of "neurons" working all together to manage a specific problem. During this study, modeling of facing parameters for GFRP using ANN is done. A back propagation (BP) multi-layer feed-forward arrangement is used for

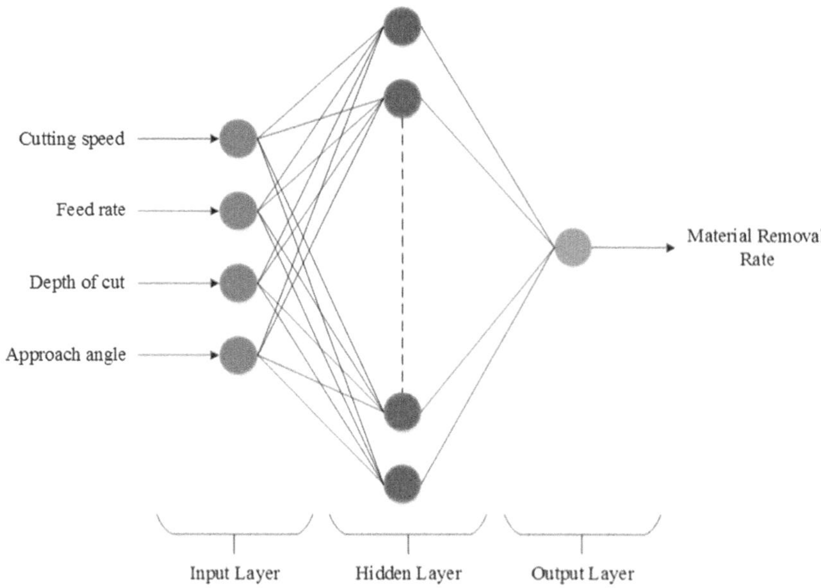


*Figure 7.5* A multi-layered artificial neural network.

this problem. Figure 7.5 shows the architecture of a BP network used for the problem. The neuron in the input layer receives input. Each input is multiplied by the connection weight. Hidden nodes sum the weighted contribution signal and apply activation function and send the output that becomes the input for an output layer. Every output neuron sums its weighted input signals and finds the output after applying an activation function, shown in Figure 7.6. Linear and tan sigmoid are the activation functions for output layer and hidden layer, respectively. Each output neuron calculates the difference between target analogues to input and output and propagates the error back and modifies the weights. The delta rule is used for updating weights. The change in weight of the neuron is given by Equation 7.4:

$$\delta w_i = \eta(t - y)x_i \tag{7.4}$$

where $\eta$ is the learning rate and $y$ is the input to the output neurons, $t$ is the objective vector, and $y$ is the output obtained by applying an activation function to $y_{in}$; here, $y_{in}$ is the summation of weighted input. The input and the corresponding output (target) are presented to the network until it learned the desired relationship. In this research work, cutting speed, feed, approach angle, and DOC are defined as input variables and material removal rate as an output parameter.

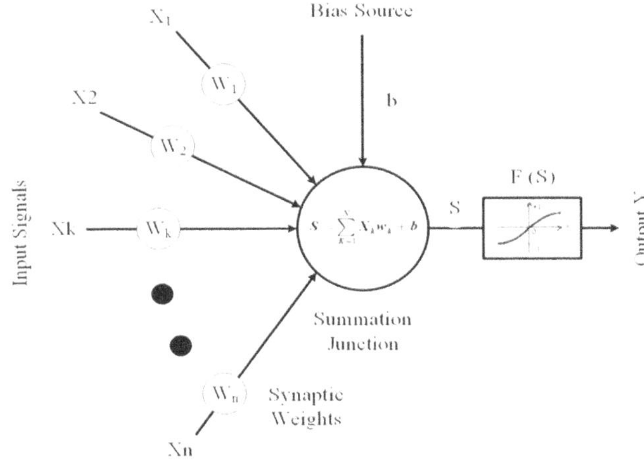


*Figure 7.6* Simple neuron model.

## 7.3 RESULTS AND DISCUSSION

The effect of explanatory parameters on MRR is found by using $L_{27}$ OA (Table 7.3). Table 7.3 shows the MRR and corresponding S/N ratio using a polycrystalline diamond cutting tool with 0.8 mm tool nose radius. The analysis is performed using an arithmetical package, Minitab –17.0, to measure the result of the machining factors on the responses. As an outcome of the ANOVA, it can be determined that cutting speed (A), feed (B), and DOC(C), have important effects on MRR, while the approach angle (D) has a refusal effect at a 95% confidence level. The effect of feed rate is extensively better than the speed and DOC. The percent contributions of machining parameters are 44.69%, 30.38%, and 18.30%, respectively, as specified in Table 7.4.

The response Table 7.5 shows is the highest contribution of feed (Δ = 0.03966) on the material removal rate, followed by speed (Δ = 0.02885) and DOC (Δ = 0.02520). Table 7.5 shows the optimum machining conditions to be A-550 rpm, B-0.1 mm/rev, C-1.0 mm, and D-75°. The response curves for the effects of four explanatory parameters for the common value of MRR and S/N ratio are plotted, as revealed in Figure 7.7(A & B).

*Table 7.4* ANOVA results for MRR

| *Variables* | *DOF* | *SS* | *MS* | *F-Value* | *P-Value* | *% contribution* | *Remarks* |
|---|---|---|---|---|---|---|---|
| Cutting velocity | 2 | 0.004815 | 0.002407 | 43.86 | 0 | 30.38 | Significant |
| Feed | 2 | 0.007084 | 0.003542 | 64.53 | 0 | 44.69 | Significant |
| DOC | 2 | 0.0029 | 0.00145 | 26.42 | 0 | 18.30 | Significant |
| Approach angle | 2 | 0.000061 | 0.00003 | 0.56 | 0.583 | 0.39 | Insignificant |
| Error | 18 | 0.000988 | 0.000055 | | | 6.23 | |
| Total | 26 | 0.015847 | | | | | |

*Table 7.5* Response table for material removal rate (Raw Data)

| | *Cutting speed (A)* | *Feed (B)* | *DOC (C)* | *Approach angle (D)* |
|---|---|---|---|---|
| L1 | 0.03964 | 0.03849 | 0.04530 | 0.05935 |
| L2 | 0.06848 | 0.05906 | 0.05993 | 0.05987 |
| L3 | 0.06759 | 0.07816 | 0.07049 | 0.05649 |
| Differences | 0.02885 | 0.03966 | 0.02520 | 0.00338 |
| Rank | 2 | 1 | 3 | 4 |

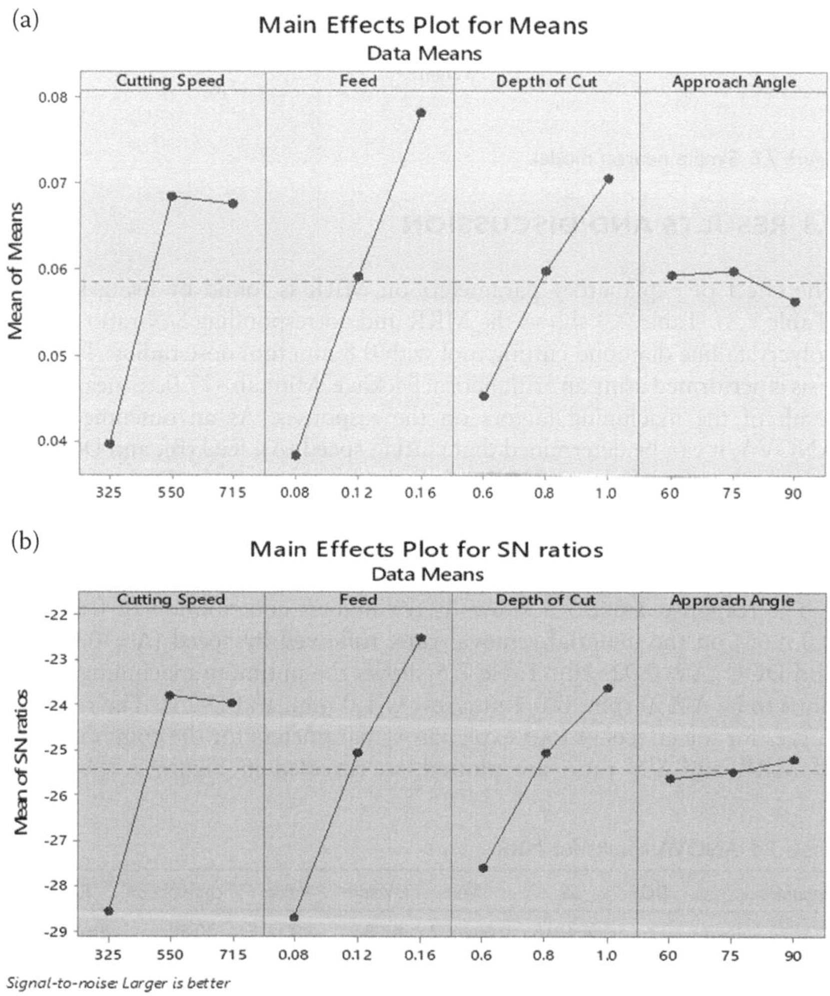


*Figure 7.7* (A) Effect of speed, feed, DOC, and approach angle on response; (B) effect of speed, feed, DOC, and approach angle on S/N ratio.

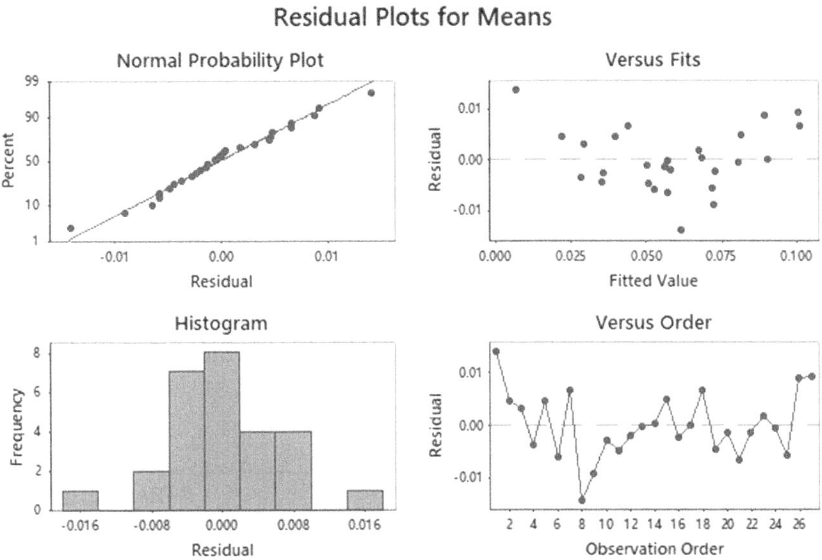


*Figure 7.8* Residual plots for means (MRR).

Figure 7.7(A) shows that the MRR increases as the feed rate and DOC increase and it increases with an increase in cutting rate and approach angle up to some limit and then it decreases. Residual plots are drawn for mean in Figure 7.8. Figure 7.8 shows that each point on the standard plot is very close to the straight line. This implies that the records are usual and a small variation from the regularity is experimental. Figure 7.8(b–d) also shows that the data does not have any distinctive structure.

### 7.3.1 Multiple Regression Analysis

The regression equation is modeled to find the connection between process parameters to assess MRR for some combination of factors levels in the given range. The efficient association between explained variables with the explanatory variables can be postulated by Equation 7.5.

$$\text{MRR} = -0.0839 + 0.000075\text{ A} + 0.4958\text{ B} + 0.0630\text{ C} - 0.000095\text{ D} \quad (7.5)$$

The regression coefficients $R^2$ and $R^2$ *(adj)* are found to be 87.33% and 85.02%, as given in Table 7.6.

*Table 7.6* Model outline

| *Model Outline* | | | |
|---|---|---|---|
| *S* | *R-sq* | *R-sq (adj)* | *R-sq (pred)* |
| 0.0095589 | 87.33% | 85.02% | 80.44% |

### 7.3.2 Artificial Neural Network

The system consists of four nodes in the input layer and one node in the output layer. Three to ten nodes are used in the intermediate layer. The optimum numbers of nodes in the intermediate layer are established. The data is normalized and then used to train the network. The parameters shown in Table 7.7 are used. A cross-validation technique is used to judge the model on unseen data. About 60% of the data is used for training, 20% for confirmation, and 20% data for test. After configuring the ANN, it is trained using the training data set and its throughput is evaluated. The training goes on until the desired outcome from the ANN is not obtained using input data sets [1996].

Figure 7.9 represents the performance plot during the training of ANN. The training is stopped when the error started rising in the validation set.

*Table 7.7* Parameters set for ANN

| | |
|---|---|
| Training function | TRAINLM |
| Learning function | LEARNGD |
| Performance function | MSE |
| No of intermediate layers | 1 |
| No. of neurons in intermediate layer | 10 |

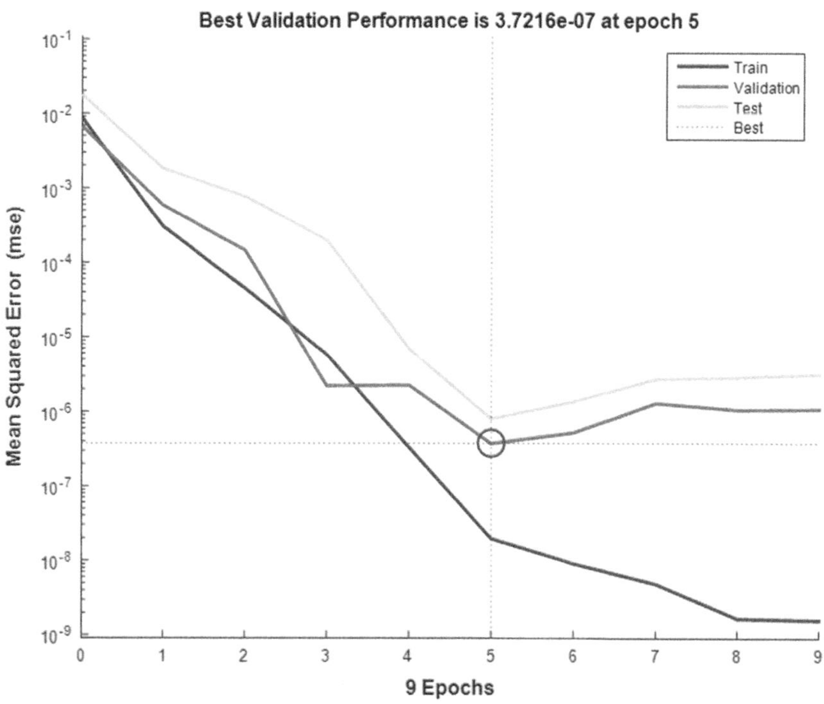


*Figure 7.9* Mean square error vs. epochs.

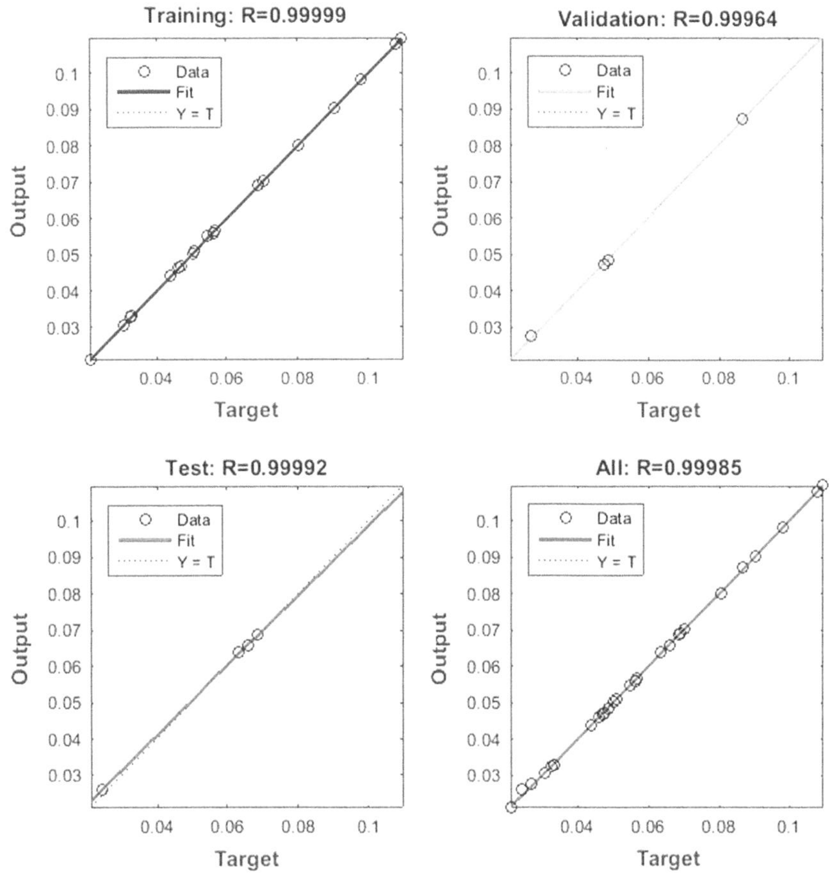


*Figure 7.10* Regression for training, validation, testing, and total data.

After obtaining the configured network, it is validated and the performance is evaluated by creating a regression plot, which shows the relationship connecting the outputs of the ANN and the experimental values. For a perfect ANN, the network-predicted values and experimental values would be accurately the same, but the connection is seldom perfect in actuality. Figure 7.10 depicts the regression plots for the developed neural network of facing operation for training, validation, and test sets. One can easily observe that the data is aligned along the line. The R value in each case is close to unity (more than 0.9). Where R = 1 represents an excellent linear relationship between predicted and experimentally obtained values. If R is near zero, then there is a non-linear connection between the predicted and experimental values. Figure 7.11(a) shows the progress of the training algorithm with respect to epochs. From the figure, it is clear that training stopped after nine epochs. Figure 7.11(b) represents the histogram plot for

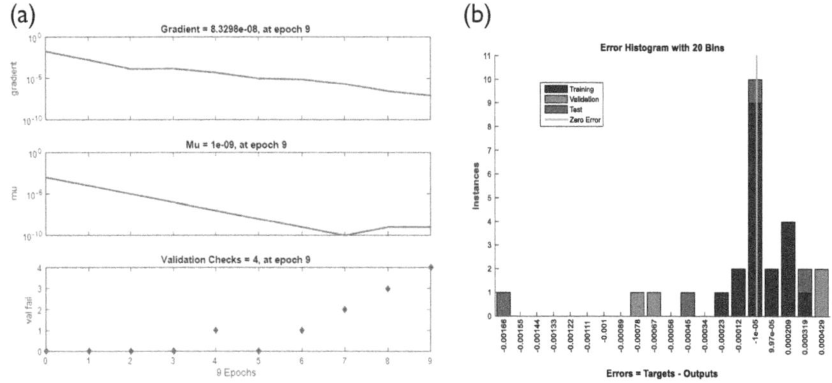


*Figure 7.11* (a) Training progress with no. of epochs, (b) bar plot of error.

errors. The predicted results obtained by the two methodologies (MRA and ANN) are compared with experimental values and are tabulated in Table 7.8. Figure 7.12 shows the experimental values and those predicted by regression and ANN. The mean absolute percentage error obtained by ANN and MRM are –13.68% and –17.90%, respectively.

*Table 7.8* Comparison of values (experimental MRM and ANN)

| *No. of experiment* | *Experimental* | *Predicted MRR by ANN (mg/s)* | *Error experimental vs. ANN* | *% Error experimental vs. ANN* | *Predicted MRR by regression (mg/s)* | *Error experimental vs. regression* | *% Error experimental vs. regression* |
|---|---|---|---|---|---|---|---|
| 1 | 0.03 | 0.02 | −0.01 | −50.00 | 0.01 | −0.02 | −200.00 |
| 2 | 0.02 | 0.03 | 0.01 | 33.33 | 0.02 | 0 | 0.00 |
| 3 | 0.04 | 0.03 | −0.01 | −33.33 | 0.03 | −0.01 | −33.33 |
| 4 | 0.05 | 0.03 | −0.02 | −66.67 | 0.03 | −0.02 | −66.67 |
| 5 | 0.05 | 0.04 | −0.01 | −25.00 | 0.04 | −0.01 | −25.00 |
| 6 | 0.05 | 0.05 | 0 | 0.00 | 0.06 | 0.01 | 16.67 |
| 7 | 0.06 | 0.05 | −0.01 | −20.00 | 0.05 | −0.01 | −20.00 |
| 8 | 0.03 | 0.05 | 0.02 | 40.00 | 0.06 | 0.03 | 50.00 |
| 9 | 0.05 | 0.06 | 0.01 | 16.67 | 0.08 | 0.03 | 37.50 |
| 10 | 0.06 | 0.03 | −0.03 | −100.00 | 0.03 | −0.03 | −100.00 |
| 11 | 0.06 | 0.05 | −0.01 | −20.00 | 0.04 | −0.02 | −50.00 |
| 12 | 0.07 | 0.06 | −0.01 | −16.67 | 0.05 | −0.02 | −40.00 |
| 13 | 0.09 | 0.06 | −0.03 | −50.00 | 0.05 | −0.04 | −80.00 |
| 14 | 0.07 | 0.07 | 0 | 0.00 | 0.06 | −0.01 | −16.67 |
| 15 | 0.09 | 0.09 | 0 | 0.00 | 0.07 | −0.02 | −28.57 |
| 16 | 0.11 | 0.07 | −0.04 | −57.14 | 0.07 | −0.04 | −57.14 |
| 17 | 0.03 | 0.09 | 0.06 | 66.67 | 0.08 | 0.05 | 62.50 |
| 18 | 0.05 | 0.11 | 0.06 | 54.55 | 0.09 | 0.04 | 44.44 |
| 19 | 0.05 | 0.03 | −0.02 | −66.67 | 0.04 | −0.01 | −25.00 |
| 20 | 0.05 | 0.04 | −0.01 | −25.00 | 0.05 | 0 | 0.00 |

*(Continued)*

*Table 7.8* (continued)

| *No. of experiment* | *Experimental* | *Predicted MRR by ANN (mg/s)* | *Error experimental vs. ANN* | *% Error experimental vs. ANN* | *Predicted MRR by regression (mg/s)* | *Error experimental vs. regression* | *% Error experimental vs. regression* |
|---|---|---|---|---|---|---|---|
| 21 | 0.07 | 0.05 | −0.02 | −40.00 | 0.06 | −0.01 | −16.67 |
| 22 | 0.08 | 0.05 | −0.03 | −60.00 | 0.06 | −0.02 | −33.33 |
| 23 | 0.07 | 0.06 | −0.01 | −16.67 | 0.07 | 0 | 0.00 |
| 24 | 0.10 | 0.08 | −0.02 | −25.00 | 0.09 | −0.01 | −11.11 |
| 25 | 0.11 | 0.07 | −0.04 | −57.14 | 0.08 | −0.03 | −37.50 |
| 26 | 0.03 | 0.09 | 0.06 | 66.67 | 0.09 | 0.06 | 66.67 |
| 27 | 0.02 | 0.11 | 0.09 | 81.82 | 0.10 | 0.08 | 80.00 |
| | | | %Mean Error | −13.68 | | %Mean Error | −17.90 |

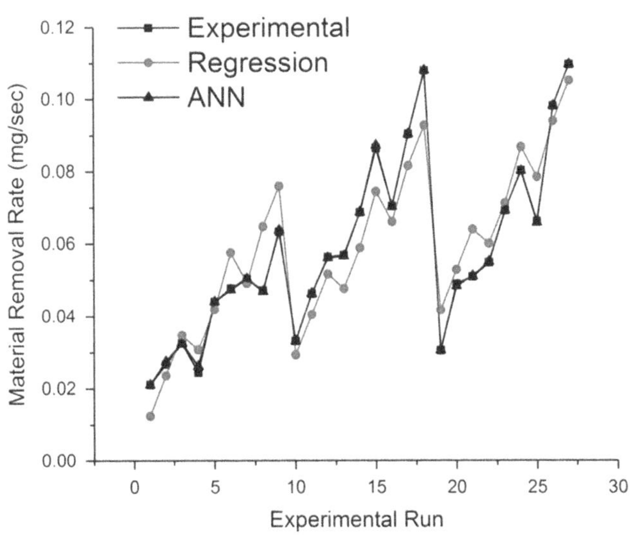


*Figure 7.12* Plot showing experimental and predicted values by MRM and ANN.

## 7.4 CONCLUSION

In the present investigation, the MRR model based on these two methods of investigation i.e. multiple regression analysis and artificial neural network is attempted effectively. The following conclusions can be drawn based on the experimental findings and the statistical analysis performed:

1. The percent contribution of feed, speed, and DOC are 44.69%, 30.38%, and 18.30%, respectively, which are significantly larger than the contribution of other factors at a 95% confidence level.

2. Feed is the most important machining parameter that affects the MRR, due to its highest percentage contribution (44.69%) amongst the process parameters.
3. The most favorable machining situation is the speed (550 rpm), feed (0.1 mm/rev), DOC (1.0 mm), and approach angle (75°).
4. $R^2$ importance for the developed model for MRR using multiple regression coefficients value is 87.33%, whereas $R^2$ obtained for ANN is more than 9.0.
5. There is good agreement between the investigational values and the predicted results. The mean absolute percentage error obtained by ANN is –13.68%, while the same for the regression model is –17.90%.
6. The developed ANN model outperforms the regression model, as seen from the results.

## REFERENCES

Arbizu I. P., Pérez C. J. L. (2003) Surface roughness prediction by factorial design of experiments in turning processes. *Journal of Materials Processing Technology*, 143–144: 390–396

Athreya, S., Venkatesh, Y.D. (2012) Application of Taguchi method for optimization of process parameters in improving the surface roughness of lathe facing operation. *International Refereed Journal of Engineering and Science*, 1(3): 13–19.

Battle M. O., (Ed.) (2008) *Introduction to Production Engineering Rio de Janeiro*. Elsevier Publisher.

Davim J. P. (2009) *Machining Composite Materials*. Wiley-ISTE, London.

Dhakad M. R., Mahajan K. A., Mitra A. C. (2017) Experimental analysis and optimization of cutting parameters for the surface roughness in the facing operation of PMMA material, *IOSR Journal of Mechanical and Civil Engineering*, 17(01): 52–60.

Dorigo M., Birattari M., Stutzle T. (2006) *IEEE Computational Intelligence Magazine* 1(4): 28–39.

Douglas Montogomery C., Elizabeth Peck A., Geoffrey Vining G. (2001) *Introduction to Linear Regression Analysis*. Arizona State University, AZ, USA.

Harvey K., Ansell Martin P. (2000) *Improved Timber Connections Using Bonded-in GFRP Rods*. Department of Materials Science and Engineering, University of Bath, Bath, UK

Haykin S. (2009) *Neural Networks and Learning Machines*. 3rd edition, Pearson Education, Inc., New Jersey.

Karakuzu J., Eberhart R. C. (1995) Particle swarm optimization. *Proceedings of IEEE international conference on Neural Networks*, 4: 1942–1948.

Karnik S. R., Gaitonde V. N., Rubio J. C., Correia A. E., Abrão A. M., Davim J. P. (2008) Delamination analysis in high speed drilling of Carbon Fiber Reinforced Plastics (CFRP) using artificial neural network model. *Materials and Design*, 29(9): 1768–1776

Keys A. A., Biajoli F. L., Mine O. M., Souza M. J. F. (2004) Exact and heuristic modeling for solving a generalization of the travelling salesman problem, In: *XXXVI Brazilian Symposium of Operational Research*, SOBRAPO, Brazil, pp. 1364–1378.

Kumar S., Meenu, Satsangi P. S. (2013) Multiple-response optimization of turning machining by the Taguchi method and the utility concept using unidirectional glass fiber-reinforced plastic composite and carbide (K10) cutting tool. *Journal of Mechanical Science and Technology*, 27(9): 2829–2837

Mayyas A., Qasaimeh A., Alzoubi K., Lu S., Hayajneh M. T., Hassan A. M. (2012) Modeling the drilling process of aluminum composites using multiple regression analysis and artificial neural networks. *Journal of Minerals and Materials Characterization and Engineering*, 11(10): 1039–1049

Meenu, Kumar S. (2013) Prediction of surface roughness in turning of UD-GFRP using Artifical Neural Network. *Mechanica Confab*, 2(3): 46–56

Myers R. H., Montgomery D. C. (1995) *Response Surface Methodology Process and Product Optimization Using Designed Experiments*. Wiley, New York, USA.

Pawan, K., Misra J. P. (2018). A surface roughness predictive model for DSS longitudinal turning operation. *DAAAM International Scientific Book* Chapter 25: 285–296.

Rojas R. (1996) *Neural Networks: A Systematic Introduction*. Springer, Germany.

Ross P. J. (1996) *Taguchi Techniques for Quality Engineering*, McGraw-Hill, New York.

Santhanakrishnan G., Krishnamurthy R., Malhotra S. K., (1989) High speed steel tool wear studies in machining of glass fibre-reinforced plastics. *Wear*, 132: 327–336.

Singh T., Kumar P., Misra J. P. (2019) Modelling of MRR during Wire-EDM of Ballistic grade alloy using Artificial Neural Network Technique. *Journal of Physics: Conference Series*, 1240: 012114. doi:10.1088/1742-6596/1240/1/012114

Suman B., Kumar P. (2006) A survey of simulated annealing as a tool for single and multiobjective optimization. *Journal of the Operational Research Society* 57(10): 1143–1160.

Uddin Md. N., Saha S. K., Hossain Md. S. (2015) Effect of lubrication condition on surface roughness in facing operation. *IJARIIE*, 1(2): 227–232.

Wang X. M., Zhang L. C. (2003) An experimental investigation into the orthogonal cutting of unidirectional fiber reinforced plastics. *International Journal Machine Tools Manufacturing*, 43(10): 1015–1022.

Zain A. M., Haron H., Sharif S. (2012) Integrated ANN-GA for estimating the minimum value for machining performance. *International Journal of Production Research*, 50(1): 191–213.

Chapter 8

# Optimization of CNC Milling Parameters for Al-CNT Composites Using an Entropy-Based Neutrosophic Grey Relational TOPSIS Method

*Sachchida Nand[1], Manvandra K Singh[1], and C M Krishna[2]*

[1]Amity University, Gwalior, Madhya Pradesh, India
[2]Maulana Azad National Institute of Technology, Bhopal, Madhya Pradesh, India

**CONTENTS**



## 8.1 INTRODUCTION

MWCNT can be developed using various techniques, out of which chemical vapour deposition technique is quite popular. CNTs are less than 100 nanometers in diameter and can be as thin as 1 or 2 nm. These materials are yet to be fully commercialized. The mechanical properties of these composites, such as tensile strength and hardness, are found to be much better than the monolithic alloy Al6061. Other mechanical properties of these composites are much greater than corresponding properties of any matrix material. In addition to these properties, there is a pressing need to investigate the machining properties of Al6061/MWCNT composites, owing to their applications in automobile and aerospace industries.

For any machining process, in addition to mechanical properties, the quality, productivity, and cost are considered to be important output parameters (criteria) at the macro level. They also become important parameters when the composites made are machined and tested for machining characteristics. Very few researchers have taken these parameters as response parameters for analysis as most of them have focused their attention on material removal rate, surface finish, cutting forces, tool wear rate, etc., at the

DOI: 10.1201/9781003258421-9

micro level. It is important to take quality, productivity, and cost as output or response parameters whenever the input parameters are tested in wider ranges of values. Quality of the machining can be interpreted in terms of surface finish and dimensional accuracy. It is important to find parameter settings that influence the quality of machining, especially in the case of new materials, such as Al6061/MWCNT composites. Input parameters such as spindle speed (in rpm), axial depth of cut (mm), feed rate (mm/rev), and radial depth of cut (also known as step-over ratio) in the case of CNC milling operations proved to have significant effects on these output parameters [1].

Many manufacturers made attempts to reduce the cost by methods such as tool selection, problem solving, etc. However, proactive strategies such as minimizing the rejected parts and unplanned downtime will help the firm in a big way to reduce the costs. After that, the focus should be on manufacturing costs and production rate. Finally, based on strong operations, manufacturers should focus on selecting tools and proper cutting conditions for optimizing the process. Manufacturers should optimize quality, productivity, and cost of a machining process at the macro level to start with and then the process can be further improved at the micro level by taking output parameters such as surface roughness, material removal rate, etc. It is believed that an increase in speed results in producing more parts and, hence, an increase in productivity. But an increase in speed also may result in faster wearing of tools, generation of heat, and vibration at the work/tool interface, and a decline in surface finish along with variation in dimensional accuracy. Hence, the speed of the rotating spindle has an influence on productivity, cost, and quality of the process. Hence, in this work, spindle speed range is taken from 2,000 to 5,000 r.p.m., varying it at three levels. Manufacturers should select the highest depth of cuts and larger feed values for each operation subject to tool clamping, workpiece fixturing, and machine tool. The ranges of feed is in 150–250 rev/min, axial depth of cut is in 0.2–0.4 mm, and radial depth of cut from 0.4–0.6 each at three levels.

It is difficult to quantify these output parameters for various combinations of input process parameters. Neutrosophic sets developed by Smarandhache [2] are widely used for handling subjective information in multi-criteria decision making (MCDM). In this research work, the combinations of input parameters are taken as various alternatives and the output parameters are taken as objectives. It is important to quantify the relationship between selected input process parameters, such as cutting speed, feed rate, and depth of cut, and output parameters in order to obtain optimum machining process parameters. For this setting, neutrosophic sets, which include indeterminacy, are better suited over fuzzy sets. Hence, in this work, neutrosophic sets are used to quantify the output variables, such as quality, productivity, and cost for a given combination of input variables.

The technique for order preference by similarity to ideal solution (TOPSIS) method is used for evaluating, prioritizing, and selecting the best alternative and proved to be a better method than many other MCDM methods [3]. In order to evaluate the weights of the objectives from the subjective data provided by experts, the information entropy method is found to be effective as it emphasizes that information related to super weight indicator is more constructive than lower indicator [4]. Thus, a novel method involving the grey relational method coupled with TOPSIS using neutrosophic sets is used in this work to find the best alternative. Literature relevant to machining process parameters and neutrosophic sets is reviewed in Section 8.2. The methodology used for finding out the best combination of input process parameters (best alternative) is explained in Section 8.3. Section 8.4 covers the results obtained and analysis made. A brief account of conclusions is given in Section 8.5.

## 8.2 LITERATURE REVIEW

Single valued neutrosophic sets (SVNSs) have been used in the past decade in the area of decision making successfully for representing imprecise, uncertain, inconsistent, and indeterminate information. Jun Ye [5] presented a decision-making method based on the correlation of intuitionistic fuzzy sets and he used a correlation coefficient of SVNSs for analysis. Nirmal N. P. et al. [6] developed a new technique for multi-attribute decision making (MADM) for converting fuzzy sets into SVNS. They used this method for the selection process of automated guided vehicle in industrial application for flexible manufacturing context. Karasan Ali et al. [7] developed a novel model with occupational health and safety as output parameters and input parameters viz., (i) probability, (ii) severity, (iii) detectability, and (iv) frequency as input process parameters. He used Pythagorean fuzzy sets in safety and critical effect analysis (SCEA) for developing the model. The development of risk assessment data was the outcome of this model. Yucesan Melih et al. [8] proposed a neutrosophic best and worst method for determining weights of failure which was carried out after pairwise comparisons. They constructed a flexible model for manufacturing industry by highlighting the contribution of knowledge about NBWM in procedure and utilization aspects.

In the area of manufacturing process design, Reddy Y. Rameswara et al. [9] used the design of experiments approach for optimizing surface roughness and MRR of super Ni 718 alloy in MCDM context for selcted machining parameters of the wire electric discharge method. Their study was based on orthogonal arrays of the Taguchi method with grey relational analysis. They used single valued neutrosophic sets for computing the weights of output parameters. A. Rosales et al. [10] proposed a method for surface quality in the face milling process and predicted surface roughness values from a wide range of cutting parameters experimentally. Benardos P.G. et al. [11] predicted surface roughness of the CNC face milling process on the basis of an artificial neural network (ANN) approach and Taguchi design of experiments (DoE) approach. In their experimentation, cutting factors in the face milling operation viz., (i) depth of cut, (ii) cutting speed, (iii) cutting forces, and (iv) feed rate, are taken into consideration. They obtained 1.86% mean squared error for surface roughness after checking of ANNs with the Levenberg-Marqardt algorithm.

Information entropy with TOPSIS was used by Huang for selection of an information system [12], Li et al. to ensure safety in coal mines [13], and Sahin and Yigider [14] for adding individual opinions of selected decision makers. A neutrosophic-based weighted distance method was used by Liu and Luo for MCDM problems [15] and they extended this method to interval-valued neutrosophic sets as well. Very few researchers have worked in relating the objectives of a manufacturing firm at the macro level with machining process parameters at the micro level using neutrosophic fuzzy sets for obtaining the best alternative of combinations of input variables using entropy-based grey relational TOPSIS.

## 8.3 METHODOLOGY

The following step-wise procedure is used for developing neutrosophic fuzzy sets for various objectives (output variables) with the help of experts' opinions, computing the weights of them, and ranking the alternatives using grey relational TOPSIS.

Step 1. To start with, identification of output variables at the macro level (firm level) and selection of input variables for a selected machining process for machining Al6061/MWCNTs in a manufacturing firm are carried out. For this study, objectives or output variables at the macro level, such as quality, productivity, and cost, are selected, as mentioned earlier. The process parameters for a milling machine viz., (i) cutting speed, (ii) feed, (iii) axial depth of cut, and (iv) radial depth of cut (step over ratio) are considered for analysis. In multi-criteria decision making (MCDM) terminology, various combinations of process parameters are considered as alternatives. In the design of experiments, they are known as treatments.

Step 2. Data are obtained from two experts, one each from the academics and the manufacturing industry who have at least 15 years of relevant experience. Input process parameters are taken at three levels each and a combination of 27 experiments are designed using a L27 array, making the number of alternatives 27. The experts are explained about the concepts of neutrosophic sets and are requested for values of membership, hesitancy and non-membership for each of the output variables for the entire set of designed experiments (alternatives). Each neutrosophic set for each of the output variable has three values corresponding to membership, hesitancy, and non-membership. The data thus can be represented by a 27 x 18 matrix for output variables in which first nine columns are used for expert-1 and the rest of the nine columns for expert-2. SVNS is explained, along with necessary notations, and used in this work, which is as follows:

A single valued neutrosophic set (SVNS) $A$ on the universe of discourse $X$ is defined as $A=\{<x,T_A(x),I_A(x),F_A(x)>,x\epsilon X\}$ where $T,I,$ $F{:}X\rightarrow[^-0,1^+]$ and $^-0\leq$ $T_A(x)+I_A(x)+F_A(x)\leq 3^+$. In this, $T_A(x)$, $I_A(x)$, *and* $F_A(x)$ represent the membership degree, hesitancy degree, and non-membership degree, respectively.

Step 3. The opinions of experts are aggregated using the union operator of neutrosophic sets in order to obtain a 27 × 9 matrix for the output variables. The union operation is performed using Equation 8.1 [16].

Let $A_m=(T_m,I_m,F_m)$ and $B_n=(T_n,I_n,F_n)$ be two SVNNs and the addition of these sets are performed as follows:

$$A_m + B_n = [Max\,(T_m,\ T_n),\ Max\,(I_m,\ I_n),\ Min\,(F_m,\ F_n)] \tag{8.1}$$

Step 4. In order to calculate the weightages of the three output variables, the information entropy method is used. The entropy method works using the information provided by data and the weights of output variables can be determined by the size of entropy measure. The values of membership degree, hesitancy degree, and non-membership degree are combined by the following operators.

The entropy measure of SVNN, $A=\{<x,T_A(x_i),I_A(x_i),F_A(x_i)>,x\epsilon X\}$ is given by

$$E_i\,(A) = 1 - \frac{\sum_{i=1}^{m} (T_A\,(x_i) + (F_A\,(x_i))|(I_A\,(x_i) - I_A^c\,(x_i))|}{m} \tag{8.2}$$

where m is the number of alternatives (rows) and $I_A^c\,(x_i)$ is a compliment of $I_A\,(x_i)$ and thus

$$I_A^c\,(x_i) = 1 - I_A\,(x_i) \tag{8.3}$$

Weights of each output parameter,

$$w_i = \frac{1 - E_i\,(A)}{\sum_{j=1}^{n} (1 - E_j\,(A))} \tag{8.4}$$

From these weights of output parameters in objective form with $w_j \leq 0$ and $\sum_{j=1}^{n} w_j = 1$ are obtained.

Step 5. Normalization using the grey relational and TOPSIS method is carried out using neutrosophic sets from the following operations. Firstly, single-valued neutrosophic relational coefficient positive (SVNRCP) $X^+$ and single-valued neutrosophic relational coefficient negative (SVNRCN) $X^-$ are found using the following procedure.

The SVNRCP $X^+$ is defined as

$X^+=\{x_1^+, x_2^+, x_3^+\}$, *where* $x_j^+ = (T_j^+, I_j^+, F_j^+,) = (max_{1\leq i<m} T_{ij}, min_{1\leq i<m} I_{ij}, min_{1\leq i<m} F_{ij}),$ *for* $j=1,2, \ldots n.$

The grey relational coefficient between $x_i$ and the SVNRCP $X^+$ on the $j^{th}$ attribute is

$$g_{ij}^+ = \frac{min_{1\leq i<m} min_{1\leq j<n}\, \Delta_{ij}^+ + \xi max_{1\leq i<m}\, max_{1\leq j<n}\, \Delta_{ij}^+}{\Delta_{ij}^+ + \xi max_{1\leq i<m}\, max_{1\leq j<n}\, \Delta_{ij}^+} \tag{8.5}$$

where $\Delta_{ij}^+ = d_N(x_{ij}, x_j^+)$, for i=1,2, … .. m, and j=1,2, … … n, $\xi = 0.5$. $d_N(x_{ij}, x_j^+)$ is computed using the following equation:

$$d_N(x_{ij}, x_j^+) = \frac{|T_{ij} - T_j^+| + |I_{ij} - I_j^+| + |F_{ij} - F_j^+|}{3} \tag{8.6}$$

From the above, SVNRCP is found by multiplying the grey relational coefficients with corresponding entropy weights of output variables using Equation 8.9.

The SVNRCN $X^-$ is defined as

$X^-=\{x_1^-, x_2^-, x_3^-\}$, *where* $x_j^- = (T_j^-, I_j^-, F_j^-,) = (min_{1\leq i<<m} T_{ij}, max_{1\leq i<<m} I_{ij}, max_{1\leq i<<m} F_{ij},$ *for* $j=1,2, \ldots n.$

Step 6. The grey relational coefficient is computed for SVNRCN. The grey relational coefficient (GRC) between $x_i$ and the SVNRCN $X^-$ on the $j^{th}$ attribute is

$$g_{ij}^- = \frac{min_{1\leq i<m} min_{1\leq j<n}\, \Delta_{ij}^- + \xi max_{1\leq i<m}\, max_{1\leq j<n}\, \Delta_{ij}^-}{\Delta_{ij}^- + \xi max_{1\leq i<m},\, max_{1\leq j<n} \Delta_{ij}^-} \tag{8.7}$$

where $\Delta_{ij}^- = d_N(x_{ij}, x_j^-)$, for i=1,2, … .. m, and j=1,2, … … n, $\xi = 0.5$. $d_N(x_{ij}, x_j^-)$ is computed using the following equation:

$$d_N(x_{ij}, x_j^-) = \frac{|T_{ij} - T_j^-| + |I_{ij} - I_j^-| + |F_{ij} - F_j^-|}{3} \quad (8.8)$$

Step 6 Compute SVNRCP and SVNRCN using Equations 8.9 and 8.10.

$$\text{SVNRCP} = \sum_{j=1}^{n} g_{ij}^{+} * w_i \text{ for j = 1, 2, ... n} \quad (8.9)$$

$$\text{SVNRCN} = \sum_{j=1}^{n} g_{ij}^{-} * w_i \text{ for j = 1, 2, ... n} \quad (8.10)$$

Compute neutrosophic relative relational degree $N_i$.

$$N_i = \frac{SVNRCP}{SVNRCP + SVNRCN} \quad \forall \text{ i = 1, 2, ... m} \quad (8.11)$$

The above 27 × 9 matrix is reduced to 27 × 3 matrix by converting the neutrosophic sets to crisp numbers.

## 8.4 RESULTS AND ANALYSIS

The designed experiments with various combinations of spindle speed (V in r.p.m.), feed (F in mm/rev), axial depth of cut (d in mm), and radial depth of cut (SR in ratio) are given in Table 8.1 along with neutrosophic sets given by first expert for the selected three output variables.

In the above table, for experiment number-1 (alternative 1), expert-1 provided three values each for each of the output variables representing membership (T), hesitancy (I), and non-membership (F) for neutrosophic sets. Similarly, the data provided by expert-2 in terms of neutrosophic sets are shown in Table 8.2.

The data provided by both experts is aggregated using an union operator, as given in Equation 8.1. The results are shown in Table 8.3 (first nine columns). Entropy calculations are performed for each row using Equation 8.2 and the results are shown in the last three columns of Table 8.3.

Table 8.4 gives results of the computations of SVNRCP row-wise. Maximum values and minimum values of T, I, and F are computed as follows:

| | | |
|---|---|---|
| $T_{max} = \max_{1 \le i \le m} (T_{ij})$; | $I_{max} = \max_{1 \le i \le m} (I_{ij})$; | $F_{max} = \max_{1 \le i \le m} (F_{ij})$; |
| $T_{min} = \min_{1 \le i \le m} (T_{ij})$; | $I_{min} = \min_{1 \le i \le m} (I_{ij})$; | $F_{min} = \min_{1 \le i \le m} (F_{ij})$; |

*Table 8.1* Neutrosophic sets given by expert-I for 27 combinations of input variables

| *Expt. no.* | *Spindle speed V* | *Feed F* | *Depth of cut D* | *Step over ratio SR* | *Expert-I* | | | | | | | | |
|---|---|---|---|---|---|---|---|---|---|---|---|---|---|
| | | | | | *Quality* | | | *Cost* | | | *Productivity* | | |
| | | | | | *T* | *I* | *F* | *T* | *I* | *F* | *T* | *I* | *F* |
| 1 | 2,000 | 150 | 0.2 | 0.5 | 0.4 | 0.2 | 0.3 | 0.7 | 0.35 | 0.25 | 0.8 | 0.25 | 0.25 |
| 2 | 2,000 | 150 | 0.3 | 0.6 | 0.35 | 0.25 | 0.3 | 0.65 | 0.4 | 0.3 | 0.7 | 0.2 | 0.3 |
| 3 | 2,000 | 150 | 0.4 | 0.4 | 0.3 | 0.15 | 0.25 | 0.2 | 0.25 | 0.2 | 0.6 | 0.25 | 0.05 |
| 4 | 2,000 | 200 | 0.2 | 0.6 | 0.35 | 0.2 | 0.4 | 0.55 | 0.4 | 0.3 | 0.7 | 0 | 0.25 |
| 5 | 2,000 | 200 | 0.3 | 0.4 | 0.3 | 0.15 | 0.25 | 0.5 | 0.3 | 0.25 | 0.6 | 0.1 | 0.2 |
| 6 | 2,000 | 200 | 0.4 | 0.5 | 0.25 | 0.15 | 0.2 | 0.45 | 0.3 | 0.25 | 0.8 | 0.05 | 0.25 |
| 7 | 2,000 | 250 | 0.2 | 0.4 | 0.45 | 0.2 | 0.4 | 0.5 | 0.25 | 0.3 | 0.5 | 0.05 | 0.3 |
| 8 | 2,000 | 250 | 0.3 | 0.5 | 0.35 | 0.2 | 0.3 | 0.45 | 0.2 | 0.25 | 0.5 | 0 | 0.2 |
| 9 | 2,000 | 250 | 0.4 | 0.6 | 0.3 | 0.25 | 0.2 | 0.4 | 0.15 | 0.2 | 0.5 | 0.15 | 0.05 |
| 10 | 3,500 | 150 | 0.2 | 0.5 | 0.5 | 0.2 | 0.2 | 0.55 | 0.4 | 0.25 | 0.55 | 0 | 0.25 |
| 11 | 3,500 | 150 | 0.3 | 0.6 | 0.5 | 0.15 | 0.15 | 0.5 | 0.3 | 0.15 | 0.55 | 0.05 | 0.1 |
| 12 | 3,500 | 150 | 0.4 | 0.4 | 0.45 | 0.15 | 0.25 | 0.55 | 0.3 | 0.35 | 0.55 | 0.05 | 0.15 |
| 13 | 3,500 | 200 | 0.2 | 0.6 | 0.35 | 0.2 | 0.1 | 0.4 | 0.4 | 0.35 | 0.5 | 0.1 | 0.1 |
| 14 | 3,500 | 200 | 0.3 | 0.4 | 0.35 | 0.2 | 0.05 | 0.35 | 0.35 | 0.3 | 0.65 | 0.05 | 0.2 |
| 15 | 3,500 | 200 | 0.4 | 0.5 | 0.3 | 0.1 | 0.15 | 0.3 | 0.25 | 0.25 | 0.6 | 0 | 0.15 |
| 16 | 3,500 | 250 | 0.2 | 0.4 | 0.45 | 0.15 | 0.25 | 0.55 | 0.3 | 0.35 | 0.7 | 0.1 | 0.25 |
| 17 | 3,500 | 250 | 0.3 | 0.5 | 0.4 | 0.1 | 0.2 | 0.5 | 0.25 | 0.25 | 0.65 | 0.15 | 0.1 |

| | | | | | | | | | | | | | |
|---|---|---|---|---|---|---|---|---|---|---|---|---|---|
| 18 | 3,500 | 250 | 0.4 | 0.6 | 0.35 | 0.15 | 0.15 | 0.45 | 0.25 | 0.35 | 0.6 | 0.05 | 0.3 |
| 19 | 5,000 | 150 | 0.2 | 0.5 | 0.7 | 0.1 | 0.2 | 0.5 | 0.2 | 0.25 | 0.65 | 0.05 | 0.2 |
| 20 | 5,000 | 150 | 0.3 | 0.6 | 0.6 | 0.05 | 0.15 | 0.45 | 0.35 | 0.3 | 0.55 | 0.2 | 0.35 |
| 21 | 5,000 | 150 | 0.4 | 0.4 | 0.5 | 0.15 | 0.2 | 0.4 | 0.25 | 0.3 | 0.5 | 0 | 0.25 |
| 22 | 5,000 | 200 | 0.2 | 0.6 | 0.55 | 0.05 | 0.1 | 0.45 | 0.3 | 0.3 | 0.6 | 0.25 | 0.4 |
| 23 | 5,000 | 200 | 0.3 | 0.4 | 0.45 | 0.1 | 0 | 0.4 | 0.1 | 0.3 | 0.1 | 0.2 | 0 |
| 24 | 5,000 | 200 | 0.4 | 0.5 | 0.45 | 0.1 | 0.05 | 0.35 | 0.15 | 0.2 | 0.1 | 0.15 | 0 |
| 25 | 5,000 | 250 | 0.2 | 0.4 | 0.5 | 0.05 | 0.05 | 0.6 | 0.1 | 0.25 | 0.1 | 0.2 | 0.05 |
| 26 | 5,000 | 250 | 0.3 | 0.5 | 0.45 | 0.1 | 0.1 | 0.55 | 0.15 | 0.2 | 0.2 | 0.1 | 0.15 |
| 27 | 5,000 | 250 | 0.4 | 0.6 | 0.4 | 0.15 | 0.15 | 0.5 | 0.2 | 0.25 | 0.25 | 0.05 | 0.25 |

*Table 8.2* Neutrosophic sets given by expert-II for 27 combinations of input variables

| *Expt. no.* | *Spindle speed V* | *Feed F* | *Depth of cut d* | *Step over ratio SR* | *Expert-2* | | | | | | | | |
|---|---|---|---|---|---|---|---|---|---|---|---|---|---|
| | | | | | *Quality* | | | *Cost* | | | *Productivity* | | |
| | | | | | *T* | *I* | *F* | *T* | *I* | *F* | *T* | *I* | *F* |
| 1 | 2,000 | 150 | 0.2 | 0.5 | 0.5 | 0.35 | 0.3 | 0.6 | 0.4 | 0.25 | 0.55 | 0.15 | 0.2 |
| 2 | 2,000 | 150 | 0.3 | 0.6 | 0.55 | 0.3 | 0.25 | 0.55 | 0.45 | 0.3 | 0.5 | 0.1 | 0.15 |
| 3 | 2,000 | 150 | 0.4 | 0.4 | 0.45 | 0.25 | 0.2 | 0.4 | 0.3 | 0.2 | 0.6 | 0.15 | 0.1 |
| 4 | 2,000 | 200 | 0.2 | 0.6 | 0.5 | 0.2 | 0.15 | 0.4 | 0.25 | 0.3 | 0.65 | 0.25 | 0.1 |
| 5 | 2,000 | 200 | 0.3 | 0.4 | 0.35 | 0.2 | 0.25 | 0.45 | 0.2 | 0.25 | 0.6 | 0.15 | 0.3 |
| 6 | 2,000 | 200 | 0.4 | 0.5 | 0.5 | 0.25 | 0.15 | 0.45 | 0.25 | 0.2 | 0.55 | 0.2 | 0.25 |
| 7 | 2,000 | 250 | 0.2 | 0.4 | 0.5 | 0.35 | 0.2 | 0.5 | 0.35 | 0.2 | 0.45 | 0.05 | 0.1 |
| 8 | 2,000 | 250 | 0.3 | 0.5 | 0.5 | 0.3 | 0.25 | 0.55 | 0.2 | 0.15 | 0.5 | 0.1 | 0.1 |
| 9 | 2,000 | 250 | 0.4 | 0.6 | 0.4 | 0.4 | 0.2 | 0.5 | 0.25 | 0.05 | 0.55 | 0.15 | 0.15 |
| 10 | 3,500 | 150 | 0.2 | 0.5 | 0.5 | 0.15 | 0.15 | 0.6 | 0.15 | 0.05 | 0.5 | 0.1 | 0.25 |
| 11 | 3,500 | 150 | 0.3 | 0.6 | 0.5 | 0.2 | 0.2 | 0.5 | 0.2 | 0.05 | 0.55 | 0.2 | 0.35 |
| 12 | 3,500 | 150 | 0.4 | 0.4 | 0.6 | 0.25 | 0.25 | 0.5 | 0.1 | 0.15 | 0.55 | 0.15 | 0.25 |
| 13 | 3,500 | 200 | 0.2 | 0.6 | 0.3 | 0.3 | 0.3 | 0.5 | 0.2 | 0.2 | 0.45 | 0.1 | 0.1 |
| 14 | 3,500 | 200 | 0.3 | 0.4 | 0.35 | 0.2 | 0.1 | 0.3 | 0.25 | 0.25 | 0.5 | 0.05 | 0.2 |
| 15 | 3,500 | 200 | 0.4 | 0.5 | 0.3 | 0.15 | 0.25 | 0.25 | 0.15 | 0.2 | 0.45 | 0.1 | 0.15 |
| 16 | 3,500 | 250 | 0.2 | 0.4 | 0.55 | 0.05 | 0.2 | 0.45 | 0.3 | 0.3 | 0.55 | 0.15 | 0.3 |
| 17 | 3,500 | 250 | 0.3 | 0.5 | 0.5 | 0.1 | 0.15 | 0.5 | 0.35 | 0.35 | 0.6 | 0.2 | 0.25 |

| | | | | | | | | | | | | | |
|---|---|---|---|---|---|---|---|---|---|---|---|---|---|
| 18 | 3,500 | 250 | 0.4 | 0.6 | 0.45 | 0.1 | 0.2 | 0.55 | 0.4 | 0.25 | 0.7 | 0.25 | 0.4 |
| 19 | 5,000 | 150 | 0.2 | 0.5 | 0.65 | 0.2 | 0.2 | 0.4 | 0.2 | 0.2 | 0.65 | 0.3 | 0.35 |
| 20 | 5,000 | 150 | 0.3 | 0.6 | 0.6 | 0.15 | 0.2 | 0.55 | 0.35 | 0.2 | 0.6 | 0.2 | 0.3 |
| 21 | 5,000 | 150 | 0.4 | 0.4 | 0.55 | 0.2 | 0.25 | 0.45 | 0.4 | 0.1 | 0.65 | 0.2 | 0.35 |
| 22 | 5,000 | 200 | 0.2 | 0.6 | 0.45 | 0.15 | 0.15 | 0.6 | 0.4 | 0.15 | 0.6 | 0.2 | 0.3 |
| 23 | 5,000 | 200 | 0.3 | 0.4 | 0.5 | 0.2 | 0.1 | 0.55 | 0.4 | 0.35 | 0.4 | 0.1 | 0.25 |
| 24 | 5,000 | 200 | 0.4 | 0.5 | 0.5 | 0.35 | 0.2 | 0.65 | 0.1 | 0.25 | 0.5 | 0.05 | 0.2 |
| 25 | 5,000 | 250 | 0.2 | 0.4 | 0.6 | 0.3 | 0.15 | 0.6 | 0 | 0.2 | 0.45 | 0 | 0.15 |
| 26 | 5,000 | 250 | 0.3 | 0.5 | 0.6 | 0.25 | 0.1 | 0.5 | 0 | 0.05 | 0.3 | 0 | 0.15 |
| 27 | 5,000 | 250 | 0.4 | 0.6 | 0.5 | 0.15 | 0.1 | 0.45 | 0.05 | 0.15 | 0.35 | 0.05 | 0.1 |

*Table 8.3* Combined sets of experts and entropy computations

| *Exp. no.* | *Quality* | | | *Cost* | | | *Productivity* | | | *Entropy computations* | | |
|---|---|---|---|---|---|---|---|---|---|---|---|---|
| | *T* | *I* | *F* | *T* | *I* | *F* | *T* | *I* | *F* | *Quality* | *Cost* | *Productivity* |
| 1 | 0.5 | 0.35 | 0.3 | 0.7 | 0.4 | 0.25 | 0.8 | 0.25 | 0.2 | 0.24 | 0.19 | 0.8 |
| 2 | 0.55 | 0.3 | 0.25 | 0.65 | 0.45 | 0.3 | 0.7 | 0.2 | 0.15 | 0.32 | 0.095 | 0.765 |
| 3 | 0.45 | 0.25 | 0.2 | 0.4 | 0.3 | 0.2 | 0.6 | 0.25 | 0.05 | 0.325 | 0.24 | 0.65 |
| 4 | 0.5 | 0.2 | 0.15 | 0.55 | 0.4 | 0.3 | 0.7 | 0.25 | 0.1 | 0.39 | 0.17 | 0.64 |
| 5 | 0.35 | 0.2 | 0.25 | 0.5 | 0.3 | 0.25 | 0.6 | 0.15 | 0.2 | 0.36 | 0.3 | 0.64 |
| 6 | 0.5 | 0.25 | 0.15 | 0.45 | 0.3 | 0.2 | 0.8 | 0.2 | 0.25 | 0.325 | 0.26 | 0.945 |
| 7 | 0.5 | 0.35 | 0.2 | 0.5 | 0.35 | 0.2 | 0.5 | 0.05 | 0.1 | 0.21 | 0.21 | 0.48 |
| 8 | 0.5 | 0.3 | 0.25 | 0.55 | 0.2 | 0.15 | 0.5 | 0.1 | 0.1 | 0.3 | 0.42 | 0.42 |
| 9 | 0.4 | 0.4 | 0.2 | 0.5 | 0.25 | 0.05 | 0.55 | 0.15 | 0.05 | 0.12 | 0.275 | 0.48 |
| 10 | 0.5 | 0.2 | 0.15 | 0.6 | 0.4 | 0.05 | 0.55 | 0.1 | 0.25 | 0.39 | 0.13 | 0.48 |
| 11 | 0.5 | 0.2 | 0.15 | 0.5 | 0.3 | 0.05 | 0.55 | 0.2 | 0.1 | 0.39 | 0.22 | 0.325 |
| 12 | 0.6 | 0.25 | 0.25 | 0.55 | 0.3 | 0.15 | 0.55 | 0.15 | 0.15 | 0.425 | 0.28 | 0.7 |
| 13 | 0.35 | 0.3 | 0.1 | 0.5 | 0.4 | 0.2 | 0.5 | 0.1 | 0.1 | 0.18 | 0.14 | 0.6 |
| 14 | 0.35 | 0.2 | 0.05 | 0.35 | 0.35 | 0.25 | 0.65 | 0.05 | 0.2 | 0.24 | 0.18 | 0.85 |
| 15 | 0.3 | 0.15 | 0.15 | 0.3 | 0.25 | 0.2 | 0.6 | 0.1 | 0.15 | 0.315 | 0.25 | 0.75 |
| 16 | 0.55 | 0.15 | 0.2 | 0.55 | 0.3 | 0.3 | 0.7 | 0.15 | 0.25 | 0.525 | 0.34 | 0.95 |
| 17 | 0.5 | 0.1 | 0.15 | 0.5 | 0.35 | 0.25 | 0.65 | 0.2 | 0.1 | 0.52 | 0.225 | 0.75 |

| | | | | | | | | | | | |
|---|---|---|---|---|---|---|---|---|---|---|---|
| 18 | 0.45 | 0.15 | 0.15 | 0.55 | 0.4 | 0.25 | 0.7 | 0.25 | 0.3 | 0.42 | 0.16 | 1 |
| 19 | 0.7 | 0.2 | 0.2 | 0.5 | 0.2 | 0.2 | 0.65 | 0.3 | 0.2 | 0.54 | 0.42 | 0.85 |
| 20 | 0.6 | 0.15 | 0.15 | 0.55 | 0.35 | 0.2 | 0.6 | 0.2 | 0.3 | 0.525 | 0.225 | 0.9 |
| 21 | 0.55 | 0.2 | 0.2 | 0.45 | 0.4 | 0.1 | 0.65 | 0.2 | 0.25 | 0.45 | 0.11 | 0.9 |
| 22 | 0.55 | 0.15 | 0.1 | 0.6 | 0.4 | 0.15 | 0.6 | 0.25 | 0.3 | 0.455 | 0.15 | 0.9 |
| 23 | 0.5 | 0.2 | 0 | 0.55 | 0.4 | 0.3 | 0.4 | 0.2 | 0 | 0.3 | 0.17 | 0.4 |
| 24 | 0.5 | 0.35 | 0.05 | 0.65 | 0.15 | 0.2 | 0.5 | 0.15 | 0 | 0.165 | 0.595 | 0.5 |
| 25 | 0.6 | 0.3 | 0.05 | 0.6 | 0.1 | 0.2 | 0.45 | 0.2 | 0.05 | 0.26 | 0.64 | 0.5 |
| 26 | 0.6 | 0.25 | 0.1 | 0.55 | 0.15 | 0.05 | 0.3 | 0.1 | 0.15 | 0.35 | 0.42 | 0.45 |
| 27 | 0.5 | 0.15 | 0.1 | 0.5 | 0.2 | 0.15 | 0.35 | 0.05 | 0.1 | 0.42 | 0.39 | 0.45 |
| | | | | | | | | | $E_i(A)$ | 0.650 | 0.733 | 0.523 |
| | | | | | | | | | 1- $E_i(A)$ | 0.350 | 0.267 | 0.477 |
| | | | | | | | | | Criteria Weights | 0.315 | 0.241 | 0.444 |

*Table 8.4* Values of $T_{max}$, $I_{max}$, and $F_{max}$ and SVNRCP computations row-wise

| | *Quality* | | | *Productivity* | | | *Cost* | | |
|---|---|---|---|---|---|---|---|---|---|
| | $T_{max}$ *0.7* | $I_{min}$ *0.1* | $F_{min}$ *0* | $T_{max}$ *0.7* | $I_{min}$ *0.1* | $F_{min}$ *0.05* | $T_{max}$ *0.8* | $I_{min}$ *0.05* | $F_{min}$ *0* |
| | $T_{min}$ *0.3* | $I_{max}$ *0.4* | $F_{max}$ *0.3* | $T_{min}$ *0.3* | $I_{max}$ *0.45* | $F_{max}$ *0.3* | $T_{min}$ *0.3* | $I_{max}$ *0.3* | $F_{max}$ *0.3* |
| Exp. no. | | | | | | | | | |
| 1 | 0.2 | 0.25 | 0.3 | 0 | 0.3 | 0.2 | 0 | 0.2 | 0.2 |
| 2 | 0.15 | 0.2 | 0.25 | 0.05 | 0.35 | 0.25 | 0.1 | 0.15 | 0.15 |
| 3 | 0.25 | 0.15 | 0.2 | 0.3 | 0.2 | 0.15 | 0.2 | 0.2 | 0.05 |
| 4 | 0.2 | 0.1 | 0.15 | 0.15 | 0.3 | 0.25 | 0.1 | 0.2 | 0.1 |
| 5 | 0.35 | 0.1 | 0.25 | 0.2 | 0.2 | 0.2 | 0.2 | 0.1 | 0.2 |
| 6 | 0.2 | 0.15 | 0.15 | 0.25 | 0.2 | 0.15 | 0 | 0.15 | 0.25 |
| 7 | 0.2 | 0.25 | 0.2 | 0.2 | 0.25 | 0.15 | 0.3 | 0 | 0.1 |
| 8 | 0.2 | 0.2 | 0.25 | 0.15 | 0.1 | 0.1 | 0.3 | 0.05 | 0.1 |
| 9 | 0.3 | 0.3 | 0.2 | 0.2 | 0.15 | 0 | 0.25 | 0.1 | 0.05 |
| 10 | 0.2 | 0.1 | 0.15 | 0.1 | 0.3 | 0 | 0.25 | 0.05 | 0.25 |
| 11 | 0.2 | 0.1 | 0.15 | 0.2 | 0.2 | 0 | 0.25 | 0.15 | 0.1 |
| 12 | 0.1 | 0.15 | 0.25 | 0.15 | 0.2 | 0.1 | 0.25 | 0.1 | 0.15 |
| 13 | 0.35 | 0.2 | 0.1 | 0.2 | 0.3 | 0.15 | 0.3 | 0.05 | 0.1 |
| 14 | 0.35 | 0.1 | 0.05 | 0.35 | 0.25 | 0.2 | 0.15 | 0 | 0.2 |
| 15 | 0.4 | 0.05 | 0.15 | 0.4 | 0.15 | 0.15 | 0.2 | 0.05 | 0.15 |
| 16 | 0.15 | 0.05 | 0.2 | 0.15 | 0.2 | 0.25 | 0.1 | 0.1 | 0.25 |
| 17 | 0.2 | 0 | 0.15 | 0.2 | 0.25 | 0.2 | 0.15 | 0.15 | 0.1 |
| 18 | 0.25 | 0.05 | 0.15 | 0.15 | 0.3 | 0.2 | 0.1 | 0.2 | 0.3 |
| 19 | 0 | 0.1 | 0.2 | 0.2 | 0.1 | 0.15 | 0.15 | 0.25 | 0.2 |
| 20 | 0.1 | 0.05 | 0.15 | 0.15 | 0.25 | 0.15 | 0.2 | 0.15 | 0.3 |
| 21 | 0.15 | 0.1 | 0.2 | 0.25 | 0.3 | 0.05 | 0.15 | 0.15 | 0.25 |
| 22 | 0.15 | 0.05 | 0.1 | 0.1 | 0.3 | 0.1 | 0.2 | 0.2 | 0.3 |
| 23 | 0.2 | 0.1 | 0 | 0.15 | 0.3 | 0.25 | 0.4 | 0.15 | 0 |
| 24 | 0.2 | 0.25 | 0.05 | 0.05 | 0.05 | 0.15 | 0.3 | 0.1 | 0 |
| 25 | 0.1 | 0.2 | 0.05 | 0.1 | 0 | 0.15 | 0.35 | 0.15 | 0.05 |
| 26 | 0.1 | 0.15 | 0.1 | 0.15 | 0.05 | 0 | 0.5 | 0.05 | 0.15 |
| 27 | 0.2 | 0.05 | 0.1 | 0.2 | 0.1 | 0.1 | 0.45 | 0 | 0.1 |

From the above values of neutrosophic sets, grey relational coefficients are computed (using Equation 8.5) after computing the hamming distance (using Equation 8.4) for each of the neutrosophic sets of output variables. From the grey relational coefficients, SVNRCP is calculated by multiplying them with corresponding weights using Equation 8.9 and the results are

*Table 8.5* Values of grey relational coefficients and SVNRCP

| *Exp. no.* | *Hamming distance 'd'* | | | *Grey relational coefficients* | | | *SVNRCP* |
|---|---|---|---|---|---|---|---|
| | *Quality* | *Productivity* | *Cost* | *Quality* | *Productivity* | *Cost* | |
| 1 | 0.24997 | 0.16665 | 0.13332 | 0.510642 | 0.648654 | 0.727279 | 0.64009 |
| 2 | 0.19998 | 0.21664 | 0.13332 | 0.58537 | 0.558143 | 0.727279 | 0.641816 |
| 3 | 0.19998 | 0.21664 | 0.14998 | 0.58537 | 0.558143 | 0.68572 | 0.623364 |
| 4 | 0.14998 | 0.23331 | 0.13332 | 0.68572 | 0.533337 | 0.727279 | 0.667448 |
| 5 | 0.23331 | 0.19998 | 0.16665 | 0.533337 | 0.58537 | 0.648654 | 0.597078 |
| 6 | 0.16665 | 0.19998 | 0.13332 | 0.648654 | 0.58537 | 0.727279 | 0.668312 |
| 7 | 0.21664 | 0.19998 | 0.13332 | 0.558143 | 0.58537 | 0.727279 | 0.639801 |
| 8 | 0.21664 | 0.11665 | 0.14998 | 0.558143 | 0.774201 | 0.68572 | 0.666857 |
| 9 | 0.26664 | 0.11665 | 0.13332 | 0.489799 | 0.774201 | 0.727279 | 0.663781 |
| 10 | 0.14998 | 0.13332 | 0.18331 | 0.68572 | 0.727279 | 0.615389 | 0.664509 |
| 11 | 0.14998 | 0.13332 | 0.16665 | 0.68572 | 0.727279 | 0.648654 | 0.679279 |
| 12 | 0.16665 | 0.14998 | 0.16665 | 0.648654 | 0.68572 | 0.648654 | 0.657587 |
| 13 | 0.21664 | 0.21664 | 0.14998 | 0.558143 | 0.558143 | 0.68572 | 0.614787 |
| 14 | 0.16665 | 0.26664 | 0.11665 | 0.648654 | 0.489799 | 0.774201 | 0.666113 |
| 15 | 0.19998 | 0.23331 | 0.13332 | 0.58537 | 0.533337 | 0.727279 | 0.635838 |
| 16 | 0.13332 | 0.19998 | 0.14998 | 0.727279 | 0.58537 | 0.68572 | 0.674627 |
| 17 | 0.11665 | 0.21664 | 0.13332 | 0.774201 | 0.558143 | 0.727279 | 0.701298 |
| 18 | 0.14998 | 0.21664 | 0.19998 | 0.68572 | 0.558143 | 0.58537 | 0.610419 |
| 19 | 0.09999 | 0.14998 | 0.19998 | 0.827595 | 0.68572 | 0.58537 | 0.685855 |
| 20 | 0.09999 | 0.18331 | 0.21664 | 0.827595 | 0.615389 | 0.558143 | 0.656817 |
| 21 | 0.14998 | 0.19998 | 0.18331 | 0.68572 | 0.58537 | 0.615389 | 0.630309 |
| 22 | 0.09999 | 0.16665 | 0.23331 | 0.827595 | 0.648654 | 0.533337 | 0.65382 |
| 23 | 0.09999 | 0.23331 | 0.18331 | 0.827595 | 0.533337 | 0.615389 | 0.662459 |
| 24 | 0.16665 | 0.08332 | 0.13332 | 0.648654 | 0.888899 | 0.727279 | 0.741463 |
| 25 | 0.11665 | 0.08332 | 0.18331 | 0.774201 | 0.888899 | 0.615389 | 0.731331 |
| 26 | 0.11665 | 0.06666 | 0.23331 | 0.774201 | 0.960012 | 0.533337 | 0.712038 |
| 27 | 0.11665 | 0.13332 | 0.18331 | 0.774201 | 0.727279 | 0.615389 | 0.692381 |

shown in Table 8.5. Similarly, SVNRCN is calculated using Equation 8.10. The results are given in Table 8.6.

From Table 8.6, the 24th, 25th, 17th and 27th sets of experiments yielded top four maximum values of N. These results are shown in Table 8.7. From these results, it may be concluded that high cutting speeds, medium values of feed, and medium values of depth of cut (both axial and radial) are preferred for best values of quality, productivity, and cost. Further experimentation at the micro level can be performed in the speed range of 3,500–5,000 r.p.m., feed range of 200–250 rev/min, axial depth of cut range of 0.3–0.4 mm, and a radial depth cut of 0.5.

*Table 8.6* Values of hamming distance, GRCs, and SVNRCN, neutrosophic relative relational degree (N)

| *Hamming distance 'd'* | | | *Grey relctional coefficients* | | | *SVNRCN* | *Exp. no.* | *N* | *Rank* |
|---|---|---|---|---|---|---|---|---|---|
| *Quality* | *Productivity* | *Cost* | *Quality* | *Productivity* | *Cost* | | | | |
| 0.083325 | 0.16665 | 0.216645 | 0.851872 | 0.621635 | 0.534894 | 0.655647 | 1 | 0.493997 | 23 |
| 0.13332 | 0.116655 | 0.216645 | 0.696985 | 0.741952 | 0.534894 | 0.635854 | 2 | 0.502333 | 19 |
| 0.13332 | 0.116655 | 0.19998 | 0.696985 | 0.741952 | 0.560987 | 0.647439 | 3 | 0.490528 | 24 |
| 0.183315 | 0.09999 | 0.216645 | 0.589756 | 0.793122 | 0.534894 | 0.614409 | 4 | 0.520689 | 13 |
| 0.09999 | 0.13332 | 0.183315 | 0.793122 | 0.696985 | 0.589756 | 0.679658 | 5 | 0.467659 | 27 |
| 0.16665 | 0.13332 | 0.216645 | 0.621635 | 0.696985 | 0.534894 | 0.601281 | 6 | 0.526398 | 10 |
| 0.116655 | 0.13332 | 0.216645 | 0.741952 | 0.696985 | 0.534894 | 0.639182 | 7 | 0.500242 | 20 |
| 0.116655 | 0.216645 | 0.19998 | 0.741952 | 0.534894 | 0.560987 | 0.611703 | 8 | 0.521569 | 12 |
| 0.06666 | 0.216645 | 0.216645 | 0.920023 | 0.534894 | 0.534894 | 0.65621 | 9 | 0.502868 | 18 |
| 0.183315 | 0.19998 | 0.16665 | 0.589756 | 0.560987 | 0.621635 | 0.596977 | 10 | 0.526767 | 9 |
| 0.183315 | 0.19998 | 0.183315 | 0.589756 | 0.560987 | 0.589756 | 0.582823 | 11 | 0.538212 | 5 |
| 0.16665 | 0.183315 | 0.183315 | 0.621635 | 0.589756 | 0.589756 | 0.599798 | 12 | 0.52298 | 11 |
| 0.116655 | 0.116655 | 0.19998 | 0.741952 | 0.741952 | 0.560987 | 0.661604 | 13 | 0.481661 | 26 |
| 0.16665 | 0.06666 | 0.23331 | 0.621635 | 0.920023 | 0.511121 | 0.644478 | 14 | 0.508254 | 16 |
| 0.13332 | 0.09999 | 0.216645 | 0.696985 | 0.793122 | 0.534894 | 0.648186 | 15 | 0.495192 | 22 |
| 0.19998 | 0.13332 | 0.19998 | 0.560987 | 0.696985 | 0.560987 | 0.593763 | 16 | 0.531877 | 8 |
| 0.216645 | 0.116655 | 0.216645 | 0.534894 | 0.741952 | 0.534894 | 0.584795 | 17 | 0.545293 | 3 |
| 0.183315 | 0.116655 | 0.149985 | 0.589756 | 0.741952 | 0.657157 | 0.656361 | 18 | 0.481866 | 25 |
| 0.23331 | 0.183315 | 0.149985 | 0.511121 | 0.589756 | 0.657157 | 0.594912 | 19 | 0.535503 | 7 |

| | | | | | | | | | |
|---|---|---|---|---|---|---|---|---|---|
| 0.23331 | 0.149985 | 0.13332 | 0.511121 | 0.657157 | 0.696985 | 0.628839 | 20 | 0.510881 | 15 |
| 0.183315 | 0.13332 | 0.16665 | 0.589756 | 0.696985 | 0.621635 | 0.629752 | 21 | 0.500221 | 21 |
| 0.23331 | 0.16665 | 0.116655 | 0.511121 | 0.621635 | 0.741952 | 0.640244 | 22 | 0.505245 | 17 |
| 0.23331 | 0.09999 | 0.16665 | 0.511121 | 0.793122 | 0.621635 | 0.628151 | 23 | 0.513291 | 14 |
| 0.16665 | 0.249975 | 0.216645 | 0.621635 | 0.489371 | 0.534894 | 0.551247 | 24 | 0.573573 | 1 |
| 0.216645 | 0.249975 | 0.16665 | 0.534894 | 0.489371 | 0.621635 | 0.562436 | 25 | 0.565272 | 2 |
| 0.216645 | 0.26664 | 0.116655 | 0.534894 | 0.469397 | 0.741952 | 0.611043 | 26 | 0.538166 | 6 |
| 0.216645 | 0.19998 | 0.16665 | 0.534894 | 0.560987 | 0.621635 | 0.579695 | 27 | 0.544292 | 4 |

*Table 8.7* N values ranked from 1 to 4

| *Rank* | *Exp. no.* | *Speed V* | *Feed F* | *Depth of cut D* | *Step-over ratio SR* | *N* |
|---|---|---|---|---|---|---|
| 1 | 24 | 5000 | 200 | 0.4 | 0.5 | 0.573573 |
| 2 | 25 | 5000 | 250 | 0.2 | 0.4 | 0.565272 |
| 3 | 17 | 3500 | 250 | 0.3 | 0.5 | 0.545293 |
| 4 | 27 | 5000 | 250 | 0.4 | 0.6 | 0.544292 |

## 8.5 CONCLUSION

In this research work, the objectives of the firm, such as optimization of quality, productivity, and cost, are taken at the macro level and they are studied for various combinations of input machining process parameters viz., (i) cutting speed, (ii) feed, (iii) radial depth of cut, and (iv) axial depth of cut. Twenty-seven combinations of experiments are considered, making the number of alternatives 27. Neutrosophic sets and expert opinions are used in order to quantify the subjective judgment of outcomes of various combinations of input machining process parameters on the selected objectives of the firm. The information entropy method is used for computing the weights of objectives, hamming distance method, and grey relational analysis are used for normalizing and ranking of alternatives is done using the TOPSIS method.

The results obtained are helpful in choosing the ranges of input process parameters for conducting experimentation at the micro level. The results indicate that the best combination of quality, productivity, and cost is obtained at a cutting speed of 5,000 r.p.m, feed of 200 rev/min, axial depth of cut of 0.4 mm, and radial depth of cut of 0.5. This study contributes to academicians and industrialists in selecting the ranges of manufacturing process parameters so that the design of experiments method can be applied in that range for further experimentation at the micro level. Further studies can be conducted for other types of machining operations, such as turning, drilling, etc., also using this expert opinion-based studies with application of neutrosophic sets.

## REFERENCES


1. V. Pare, G. Agnihotri, and C. M. Krishna. (2016) Selection of optimum process parameters in high-speed CNC end-milling of composite materials using meta heuristic techniques—A comparative study. *Strojniski Vestnik-Journal of Mechanical Engineering*, 61(3), 176–187.
2. F. Smarandhache. (1999) *A Unifying Field in Logics. A Neutrosophy: Neutrosophic Probability, Sets, and Logic*. American Research Press, Rehoboth.

3. E. Roszkowska. (2011) Multicriteria decision making models by applying the TOPSIS method to crisp and interval data. *Multiple Criteria Decision Making*, 6(1), 200–230.
4. R. Kumar, S. Singh, P. S. Bilga, Jatin, J. Singh, S. Singh, M. L. Scutaru, and C. I. Pruncu. (2021) Revealing the benefits of entropy weights method for multi-objective optimization in machining operations: A critical review. *Journal of Materials Research and Technology*, 10, 1471–1492, ISSN 2238-7854, 10.1016/j.jmrt.2020.12.114.
5. J. Ye. (2013) Multicriteria decision-making method using the correlation coefficient under single-valued neutrosophic environment. *International Journal of General Systems*, 42(4), 386–394.
6. N. P. Nirmal and M. G. Bhatt. (2018) Development of fuzzy-single valued neutrosophic madm technique to improve performance in manufacturing and supply chain functions. *Fuzzy Multi-criteria Decision-Making Using Neutrosophic Sets*, pp 711–729.
7. K. Ali, I. Esra, C. Selcuk, and K. Cengiz. (2018) A new risk assessment approach: Safety and Critical Effect Analysis (SCEA) and its extension with Pythagorean fuzzy sets. *Safety Science*, 108, October 2018, 173–187.
8. M. Yucesan and M. Gul. (2018) Failure prioritization and control using the neutrosophic best and worst method. *Granular Computing*, 6, 435–449.
9. R. Y. Rameswara and R. B. Chandra Mohana. (2018) Optimization of WEDM parameters for SUPER Ni 718 using GRA with neutrosophic sets. *International Journal of Applied Engineering Research*, 13(12), 10924–10930.
10. A. Rosales, A. Vizán, E. Diez, and A. Alanís. (2010) Prediction of surface roughness by registering cutting forces in the face milling process. *European Journal of Scientific Research*, ISSN 1450-216X, 41(2), 228–237.
11. P. G. Benardos and G. C. Vosniakos. (2002) Prediction of surface roughness in CNC face milling using neural networks and Taguchi's design of experiments. *Robotics and Computer Integrated Manufacturing*, 18, 343–354.
12. J. Huang. (2008) Combining entropy weight and TOPSIS method for information system selection. *2008 IEEE Conference on Cybernetics and Intelligent Systems*, IEEE 2008.
13. A. Li, J. Zhao, Z. Gong, and F. Lin (2016). Optimal selection of cutting tool materials based on multi-criteria decision-making methods in machining Al-Si piston alloy. *International Journal of Advanced Manufacturing Technology*, 86(1–4), 1055–1062.
14. R. Sahin and M. Yigider. (2016) A multi-criteria neutrosophic group decision making method based on TOPSIS for supplier selection. *Applied Mathematics and Information Sciences*, 10(5), 1843–1852.
15. C. Liu and Y. Luo. (2016) The weighted distance measure-based method to neutrosophic multi-attribute decision making. *Mathematical Problems in Engineering*, 2016.
16. H. Wang, F. Smarandache, Y. Zhang, and R. Sunderraman. (2010) Single valued neutrosophic sets. *Review of the Air Force Academy*, 17, 10–14.

Chapter 9

# Experimental Investigation of EDM Potential to Machine AISI 202 Using a Copper-Alloy Electrode and Its Modeling by an Artificial Neural Network

*Subhash Singh*[1] *and Girija Nandan Arka*[2]

[1]Department of Mechanical and Automation Engineering, Indira Gandhi Delhi Technical University for Women, New Delhi, India

[2]Department of Production and Industrial Engineering, National Institute of Technology (NIT), Jamshedpur, Jharkhand, India

## CONTENTS



## 9.1 INTRODUCTION

Machining of hard materials always radiates productive machining techniques to produce economical components with quality assurance. However, conventional methods failed to process, owing to tool material constraints. Nontraditional machining processes shine as alternative processes to machine such hard materials. EDM classified under nontraditional machining retains a huge potential for the processing of wider materials irrespective of hardness property. Moreover, complex shapes could be portrayed over work material make the EDM machine more versatile and robust relative to alternative

DOI: 10.1201/9781003258421-10

nontraditional machining process. Since material removal from the parent material is associating with melting and vaporizing of an exposed work area rather have physical contact. It establishes a controlled pulse spark to melt the exposed work area empowered by a pulse generator whose function is to convert continuous AC to pulse DC [1].

EDM offers a better alternative solution to do machining of high hardness materials that finds vast applications in aerospace, medical devices, electronics, automotive, and optics. Since different materials require different machining conditions of electrical energy variables to get thermal erosion and because of that electrical process parameters required to be optimized [2]. During machining there is always an optimum gap maintained between electrode and workpiece and is monitored by a servo controller, thus eliminating possibilities of mechanical stress or any type of vibration [3]. Because of its unconventional nature of doing machining, this can be used in many highly demanded application materials like ceramics, heat-treated materials, case-hardened materials, and composite [4]. This also can be applied to machine extrusion and forging dies along with other operations like drilling, milling, and grinding.

Every machine has its advantages and limitations and hence EDM also has some limitations since spark erosions take place in both the electrode and workpiece sides. Hence, tool wear taking place as been used continuously for a long time. Therefore, it hampers machining and leads to defect productions. Hence, research is needed to optimize machining parameters to prevail over, and improve, the machining performance. The EDM tool should have good machinability, electrically conductive, thermally conductive, and cost effective [5]. By considering all relevant priorities, copper or graphite tools are most commonly used in EDM. This is observed that tool wear is one of the failure conditions while doing machining. Many researchers tried to develop technology and tested to minimize tool wear rate. Khanra et al. experimentally investigated machining performance by adopting a "$ZrB_2$-Cu" tool and recorded the least tool or electrode wear rate as compared to a pure copper tool [6]. Novel researchers, Estebean et al. experimentally investigated ceramic workpiece material and found excellent machining with a desired dimensional accuracy [7]. Another research team, Kiyak et al., attempted to analyze the effects of process parameters to machine tool steel with a cupper electrode. Based on their experimental result, they gave the conclusion that input current I and pulse on ton contribute the most towards a primary surface texture of copper tools as well as for the tool steel [8]. Some researchers developed a $ZrB_2$-CuNi tool and narrated its vast applications [9]. Review of EDM manifests that several efforts have attempted to bring down the TWR minimum and maximize MRR [10]. Fonesca et al. were able draw a conclusion that deionization time of dielectric fluid during $T_{off}$ has a significant effect on wear rate of the tool as well as for the removal rate of material [11]. Another researcher team, Tang et al., experimentally investigated and optimized EDM parameters to machine a stainless steel job and stated a conclusion that the machining input parameter current I and $T_{on}$ (Pulse on) are the two machining parameters that influence a lot to maximize MRR and damaging tool or electrode [12]. Researcher team Tsai et al. had worked on a Cr/Cu composite-based tool. Based on their result, they have arrived at a conclusion that an increase in MRR is reported when the electrode connected is negative but have been detrimental to the interests of the surface finish. [13]. A researcher team, Lee et al., had investigated the interests of surface cracks and applied a full factorial design approach to get the significant effects of EDM parameters [14]. Statistical analysis of he wear rate of the tool examined by the researcher team concluded that the machining parameter current I and $T_{on}$, known as "pulse on time," have significant effects [15]. The research team Haron et al. experimentally studied two different tool electrodes: copper and graphite. They checked individual tool performance on EDM and found the copper tool superior to that of graphite while machining the tool steel alloy [16]. The research team Hu et al. conducted an experiment on ceramic material and concluded that "melting and decomposition" are the two important reasons for material removal [17]. Muttamara et al. had marked the effect of material of electrodes to machine aluminium in EDM [18]. Bhaumik M. et al. attempted a cryotreated tungsten carbide electrode and sic powder on EDM performance of AISI304 and they found improved machining efficiency with a combinational effect of a cryotreated electrode with sic powder [19]. When sic abrasive powder is mixed with dielectric fluid on EDM, there were improved surface characteristics of H-11 die steel [20]. However, a gap voltage of an EDM-machined parameter found adverse effects to MRR while machining EN19 steel [21]. To improve surface quality air and argon gas flushing found more uniform recast layer with absence of potholes and cracks on Ti6Al4V alloy [22].

It has been observed that several efforts and techniques have been applied to bring down the TWR minimum and improve the effectiveness of productive efforts of EDM. But from a literature survey, EDM machining found lacuna to do machining of stainless steel of AISI 202 grade combined with electrode Cu-Cr-Zr alloy. Along with this, RSM technology on EDM machining has never been attempted before. Therefore, in this research, machining strategy has been made to fill the research gap of EDM parametric retrospect against stainless steel of AISI 202 grade specimen. Further, ANN has been introduced to predict the experimental result. The results indicate that ANN prediction is well aligned with experimental results.

## 9.2 EXPERIMENTAL DETAILS

### 9.2.1 Details of the Tool Electrode

The copper-based alloy material copper-chromium-zirconium (Cu-Cr-Zn) has been employed for making the electrode. Since Zn improves the strength

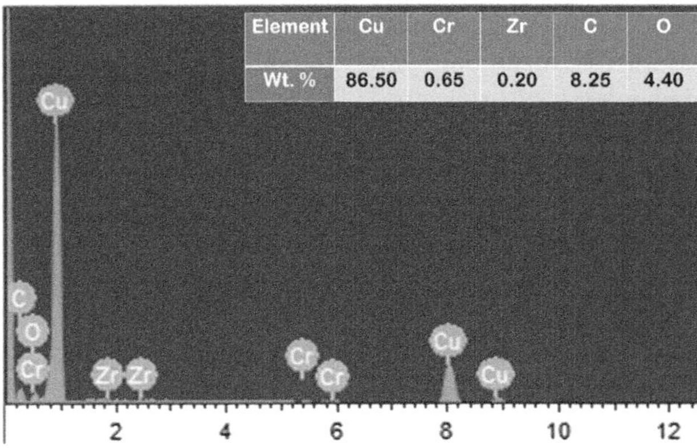


*Figure 9.1* Depicting the composition of Cu-Cr-Zr electrode material and validating the genuineness of electrode material by EDS mapping.

and retains electrical conductivity at a high temperature, Cu and Cr add value to electrical conductivity and mechanical strength, respectively; thus, to catch all those indispensable properties, Cu-Cr-Zn has been employed. The details of Cu-Cr-Zn are illustrated in Figure 9.1.

### 9.2.2 Workpiece Details

AISI 202 grade stainless steel was purchased for the experimentation, since it has wider applications in home appliances, aviation, medical, and surgical instruments, etc. The work material AISI 202 grade of stainless-steel chemical composition and energy dispersive x-ray spectroscopy (EDS) mapping is presented in Figure 9.2.

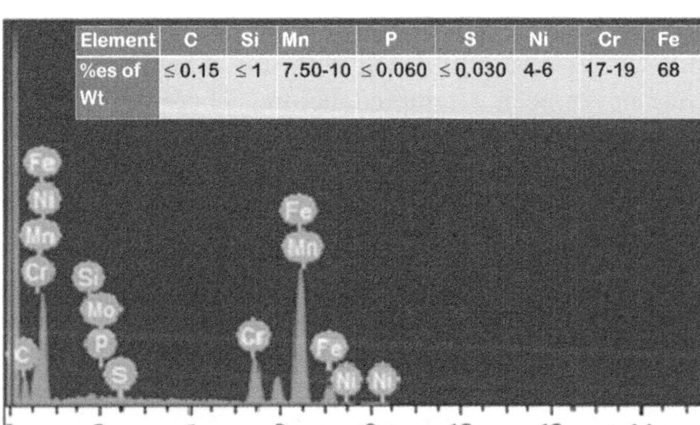


*Figure 9.2* Details of AISI 202 grade including composition and EDS mapping.

### 9.2.3 Design of Experiment

The RSM (response surface methodology) technique was studied and applied to design the experiment and optimized EDM machining parameters to get the best result. This is because RSM has a statistical algorithm that can convert a response into a mathematical model and can be used to analyze the problem with the least number of experimental runs without damaging accuracy [23]. Here, a RSM $2^k$ factorial design is adopted to design the experimental run, record results, and compute a second-order polynomial regression equation for MRR and EWR [24]. The RSM objective is to analyze the role of the EDM process parameters against various EDM output responses [25]. RSM helps to find the significance effect of EDM process parameters towards responses. A basic second-order polynomial regression equation can be of following:

$$Y = \beta_0 + \sum_{i=1}^{n} \beta_i X_i + \sum_{i=1}^{n} \beta_{ii} X_i^2 + \sum_{i=1}^{n} \beta_{ij} X_i X_j \ldots\ldots \tag{9.1}$$

where Y is the response that can be MRR and EWR, constants $\beta_o$, $\beta_i$, $\beta_{ii}$, $\beta_{ij}$ represent second-order regression coefficients, and $X_i$ (i varies from 1, 2, 3 ... n) is the quantitative machining variable. Three experimental machining process parameters with three levels are selected based on best performances, represented in Table 9.1. These parameters were taken based on literature to acquire the best experimental result.

The output parameters: MRR and TWR are quantified using the following equation presented in Equations (9.2) and (9.3). The weight of the work and tool materials for each run were measured by using a weighing machine:

$$\text{MRR} = \frac{\text{premachining workpiece weight} - \text{postmachining workpiece weight}}{\text{Time of machining}} \tag{9.2}$$

$$\text{TWR} = \frac{\text{premachining electrode weight} - \text{postmachining electrode weight}}{\text{Time of machining}} \tag{9.3}$$

*Table 9.1* Process parameters with their level addressed for the experimental investigation

| *Parameters* | *Pulse current (A)* | *Pulse off-time (μs)* | *Pulse on-time (μs)* |
|---|---|---|---|
| High level | 9 | 9 | 75 |
| Medium level | 7 | 7 | 65 |
| Low level | 5 | 5 | 55 |

## 9.3 CHARACTERIZATION

Herein, a scanning electron microscope (SEM) was incorporated to study the physical behavior of processed material in a micron level and can detect any unexpected aspect of defect so that it can be avoided further. SEM micrographs with SEM LEO 435 VP, resolution of 6 nm VP, and 4 nm HV at various magnification ranges from 500X to 6.00KX were used.

## 9.4 RESULTS AND DISCUSSION

All of the precise data for each experiment was brought together in one edition and was mathematically applied to get the values of the electrode or tool erosion rate and MRR. All machining data and calculated response values are presented in Table 9.2.

After getting machining data, the optimization technique was executed to get optimized EDM process parameters, setting a combination that satisfied industrial interests of least electrode or tool wear rate and highest material

*Table 9.2* Experimental result

| *Run no.* | *I in A* | *$T_{off}$ in μs* | *$T_{on}$ in μs* | *Experimental result in mg/min* | | *ANN prediction* |
|---|---|---|---|---|---|---|
| | | | | *MRR* | *TWR* | *MRR* |
| 1 | 5 | 9 | 75 | 0.00155 | 0.000165 | 0.001550003 |
| 2 | 7 | 7 | 55 | 0.01847 | 0.000204 | 0.01847 |
| 3 | 7 | 5 | 65 | 0.013095 | 0.000175 | 0.013095001 |
| 4 | 5 | 5 | 55 | 0.00198 | 0.000163 | 0.000282977 |
| 5 | 7 | 7 | 65 | 0.012415 | 0.000179 | 0.0141875 |
| 6 | 9 | 5 | 75 | 0.0342 | 0.000521 | 0.0342 |
| 7 | 7 | 7 | 75 | 0.00669 | 0.000273 | 0.006689997 |
| 8 | 5 | 5 | 75 | 0.001495 | 0.000168 | 0.001495006 |
| 9 | 7 | 7 | 65 | 0.015095 | 0.000181 | 0.0141875 |
| 10 | 9 | 7 | 65 | 0.045095 | 0.000286 | 0.040419053 |
| 11 | 7 | 7 | 65 | 0.01382 | 0.000184 | 0.0141875 |
| 12 | 5 | 9 | 55 | 0.001205 | 0.000155 | 0.001205001 |
| 13 | 9 | 9 | 75 | 0.00264 | 0.000413 | 0.01202863 |
| 14 | 7 | 7 | 65 | 0.013985 | 0.000187 | 0.0141875 |
| 15 | 9 | 5 | 55 | 0.048175 | 0.000372 | 0.048175 |
| 16 | 7 | 9 | 65 | 0.013375 | 0.000215 | 0.013375 |
| 17 | 7 | 7 | 65 | 0.01385 | 0.000188 | 0.0141875 |
| 18 | 5 | 7 | 65 | 0.001665 | 0.000162 | 0.001664997 |
| 19 | 9 | 9 | 55 | 0.04203 | 0.000335 | 0.048406743 |
| 20 | 7 | 7 | 65 | 0.014535 | 0.000192 | 0.0141875 |

*Table 9.3* Optimized parameters

| *Process parameters* | *Optimized controllable figure value* | |
|---|---|---|
| | *MRR* | *TWR* |
| I (A) | 9 | 5 |
| Ton (μs) | 75 | 65 |
| Toff (μs) | 9 | 7 |

removal rate of AISI 202 stainless steel. Table 9.3 presents optimal EDM machining parameters to get the best responses.

### 9.4.1 Mathematical Representation of Different Responses

By using RSM, all experimental runs have been performed and the collected data applied in MINITAB 17 software. MINITAB 17 is a statistical software with RSM methodology to get the best experimental solution. This software solution can help to perform machining analysis. By applying the RSM approach, the different responses can be converted into a mathematical model from which significant effects of input parameters can be identified. A regression equation was derived from the RSM statistical approach and presented bellow in Equations (9.4) and (9.5) for EWR and MRR:

$$\begin{aligned} \mathrm{TWR} = {} & 0.000185 + 0.000111\mathrm{I} + 0.000031\mathrm{T_{on}} - 0.000012\mathrm{T_{off}} \\ & + 0.000039\mathrm{I} \times \mathrm{I} + 0.000053\mathrm{T_{on}} \times \mathrm{T_{on}} + 0.000010\,\mathrm{T_{off}} \times \mathrm{T_{off}} \\ & + 0.000026\mathrm{I} \times \mathrm{T_{on}} - 0.000017\,\mathrm{I} \times \mathrm{T_{off}} - 0.000008\mathrm{T_{on}} \times \mathrm{T_{off}}. \end{aligned} \tag{9.4}$$

$$\begin{aligned} \mathrm{MRR} = {} & 0.01463 + 0.01642\mathrm{I} - 0.00652\,\mathrm{T_{on}} - 0.00381\,\mathrm{T_{off}} + 0.00774\mathrm{I} \times \mathrm{I} \\ & - 0.00306\,\mathrm{T_{on}} \times \mathrm{T_{on}} - 0.00240\,\mathrm{T_{off}} \times \mathrm{T_{off}} - 0.00665\mathrm{I} \times \mathrm{T_{on}} \\ & - 0.00462\mathrm{I} \times \mathrm{T_{off}} + 0.00308\mathrm{T_{on}} \times \mathrm{T_{off}}. \end{aligned} \tag{9.5}$$

### 9.4.2 Analysis of Variance

Analysis of variance, commonly known as ANOVA, uses a statistical approach to find the sum of squares of variances (SS), adj. mean square (Adj. MS), and F value calculated [26]. ANOVA is performed to know the significant effect of paramount machining parameters on responses so that further study can be controlled and improved. In this work, the F value is calculated by dividing the adjacent mean square of a variable into the adj. mean square of error and *p*-value computed for TWR at a 90% confidence

*Table 9.4* Significant parameters revealed from ANOVA for TWR

| *Source* | *DOF* | *Adj. sum of the square* | *Adj. mean of the squares* | *F calculated* | *P-value* | |
|---|---|---|---|---|---|---|
| Linear | 3 | 0.000002 | 0 | 47.59 | 0.000 | Significant |
| Square | 3 | 0 | 0 | 13.94 | 0.001 | Significant |
| 2-way interaction | 3 | 0 | 0 | 2.96 | 0.084 | Non-significant |
| Lack of fit | 5 | 0 | 0 | 81.41 | 0.000 | Significant |
| Pure error | 5 | 0 | 0 | | | |
| Total | 19 | 0 | | | | $R^2$ = 95.09% |
| | | | | | | $R^2$ (adj) = 90.66% |

*Table 9.5* Significant parameters revealed from ANOVA for MRR

| *Source* | *DOF* | *Adj. SS* | *Adj. MS* | *F calculated* | *P-value* | |
|---|---|---|---|---|---|---|
| Linear | 3 | 0.003267 | 0.001089 | 51.86 | 0.000 | Significant |
| Square | 3 | 0.000166 | 0.000055 | 2.63 | 0.108 | Non-significant |
| 2-way interaction | 3 | 0.000600 | 0.000200 | 9.52 | 0.003 | Significant |
| Lack of fit | 5 | 0.000206 | 0.000041 | 51.00 | 0.000 | Significant |
| Pure error | 5 | 0.000004 | 0.000001 | | | |
| Total | 19 | 0.004242 | | | | $R^2$ = 95.05% |
| | | | | | | $R^2$(adj) = 90.60% |

level, presented in Table 9.4. Similarly, ANOVA is obtained for MRR, presented in Table 9.5. Since $R^2$ and adjusted $R^2$ value comes to 90 to 95%, therefore the experiment performed is satisfied. Here it shows each machining parameter significantly contributes toward responses of MRR and TWR, respectively, linearly as well as square level but not contributed by a combined or interaction effect.

### 9.4.3 EDM Machining Parameters Influence on MRR

Studied the effect of EDM process parameter pulse current I, pulse off-time $T_{off}$, pulse on-time $T_{on}$ towards improvement of MRR performance. By using statistical software, a 3D surface plot has been generated for analyzing the silent feature of machining parameters, presented in Figure 9.3. MRR is directly proportional to pulse current, clearly observed in Figure 9.3. A significant MRR is detected at a high pulse current with a high pulse on-time. A similar trend was observed for pulse off-time, depicted in Figure 9.3(b). This depicts that as $T_{off}$ increases with respect to pulse current; more time is invested for the flushing of debris, which encourages a debris-free surface for next-step machining. From Figure 9.3(c), it has been

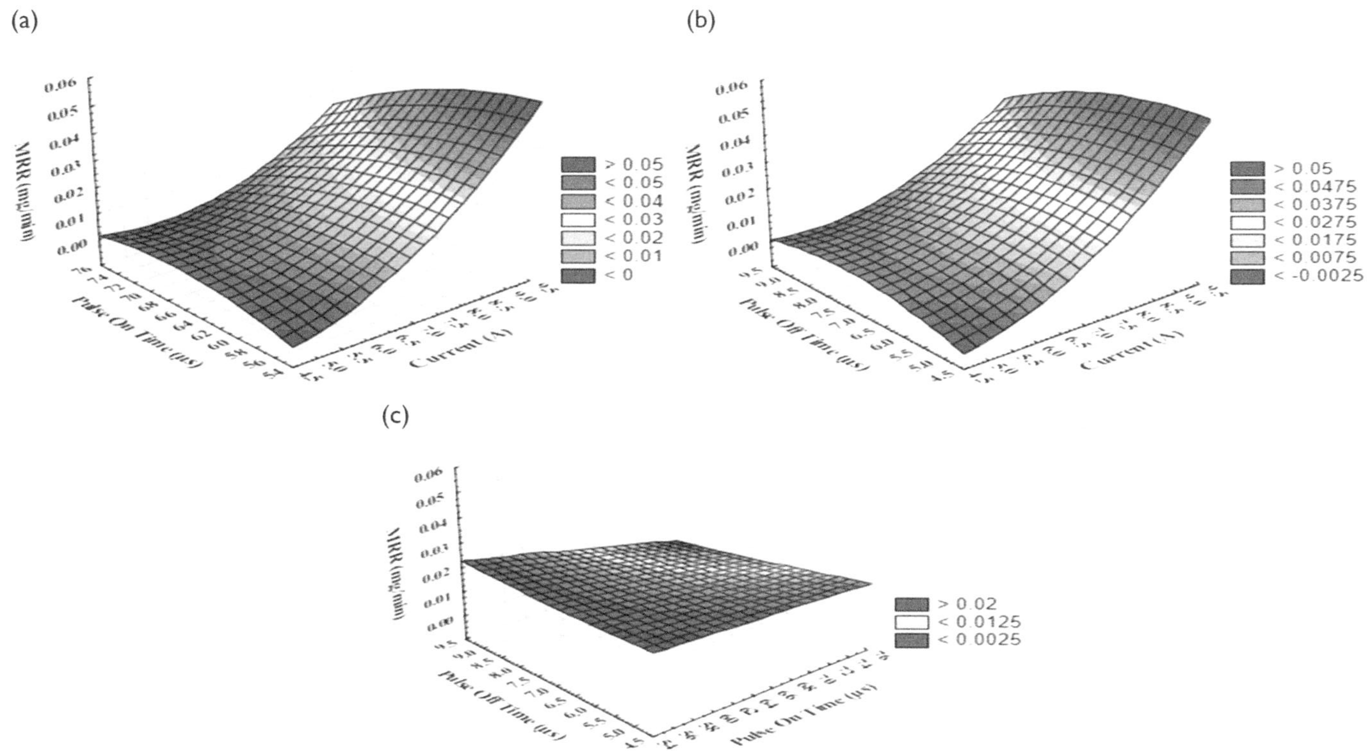


*Figure 9.3* Contribution towards MRR: (a) Influence of pulse current and pulse on-time, (b) Influence of current and pulse off-time, (c) Influence of pulse on-time and pulse off-time.

concluded that low pulse off-time disheartens MRR and this phenomenon may occur due to inefficient removal of debris that obstructs the machining process significantly. Thus, the remaining deposited sludge is exposed to the next-step pulse and obstructs the transit of heat originated from a pulsed spark and is unable to transfer a hit to the fresh work material consequence MRR reduction. Hence, high pulse off-time encourages flourishing removal of sludges from the machining zone and adequate heat transfer to the fresh work area and MRR improve significantly. This study portrayed that the $T_{off}$ duration has significant effects on the removal of material.

### 9.4.4 Study of the Surface Alteration

An EDM machined surface and tool surface are investigated by SEM. Since the machined surface undergoes various temperature changes such as getting heated by sparking and getting cooled by dielectric flushing, hence the machined surface is investigated by SEM for a detailed study at optimum controllable parameters. The machined surface is obtained at a machined parameter: pulse current 9 A, pulse on-time 75 μs, and pulse off-time 9 μs, respectively, was studied by SEM. Similarly, the electrode surface was investigated at an optimum parametric setting: pulse current 5 A, pulse on-time 65 μs, and pulse off-time 7 μs, respectively. The surface generated by an optimum setting is illustrated in Figure 9.4.

Remarkable changes on a machined surface have been reported froman SEM figure for workpiece material. The presence of micro-cracks and a recast layer are clearly visible in Figure 9.4(b). This physical alteration of the machined surface that appeared as micro-cracks shows that a greater amount of thermal stress developed due to a high temperature environment and sudden drop by quenching. The thermal stress generated by the temperature change at a local area led to forming micro-cracks. Moreover, the presence of a recast layer depicts the debris obtained due to melted and burned material drop that is unable to be flushed from the machining zone.

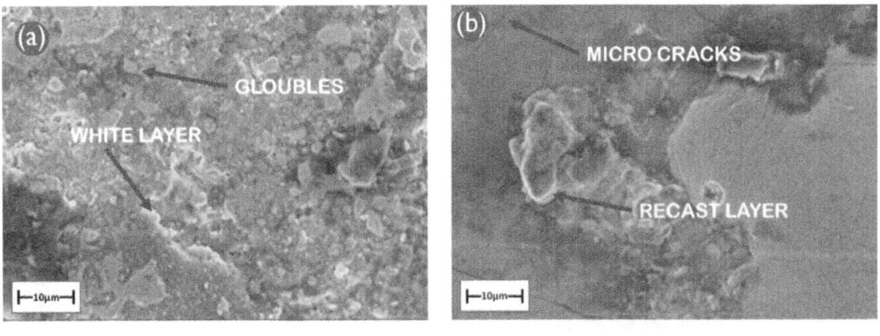


*Figure 9.4* Post-machining analysis of a tool surface and AISI 202 grade stainless steel at an optimum setting: (a) 5 A pulse current effect on tool surface, (b) 9 A pulse current effect on AISI 202 grade stainless steel.

Thus, flushing pressure could be incorporated to overpower the recast deposition. Moreover, micro-cracks developed due to increased residual stresses resulting from non-homogeneities within a white layer [27]. The melted materials were quenched by dielectric fluid forming debris deposited over a machined surface, forming a white layer. The microstructure of bulk material does not affect EDM machining [28].

The tool surface was embossed with globules and a white layer, as shown in Figure 9.4(a). The discussion agrees with the practical change. The deposited globule percentages are more; it protects the tool surface from thermal erosion; hence, electrode wear rate decreases significantly. The presence of a white layer also helps to endure the thermal erosion.

## 9.5 ARTIFICIAL NEURAL NETWORK (ANN)

An ANN is an algorithm inspired from an animal brain working principle to solve linear and nonlinear engineering problems very effectively. ANN is a computing system has an input layer and an output layer with a number of hidden layers connected with a number of neurons (processing elements). The parallel architecture of ANN enables a fast computing process compared to other conventional technique as all the neurons in between all the layers work simultaneously [29]. Due to the complex nature of EDM machining process, it is very tough to accurately predict the machining process and it gets easier by ANN modeling. Since ANN is superior to RSM, therefore, in this present work EDM modeling with a feed-forward back propagation neural network with two layers and 20 neurons is adopted using a MATLAB ANN tool for MRR [30]. The optimal structure is created by the trial-and-error method by changing the number of neurons and finding the best fit with a minimum MSE and maximum R value with more than a 95% confidence level, depicted in Figure 9.5 for MRR. The predicted result is compared to the experimental result listed in Table 9.4. The developed model is well trained and gave an average error of 0.05%. Weights were reassigned to reduce error through back propagation during training of the data using Equation (9.6), where O represents output, T represents target, and P represents input patterns, respectively. The Bayesian regularization (BR) algorithm was incorporated for the training of the data and a log sigmoid activation function was introduced to both the input and output layers. A set of experiments was conducted and their results were taken for training of the data. Figure 9.6 depicts a comparison between experimental MRR with ANN-predicted MRR.

$$Error = 0.5 \times \sum_{j=1}^{P} \|O_j - T_j\|^2 \qquad (9.6)$$

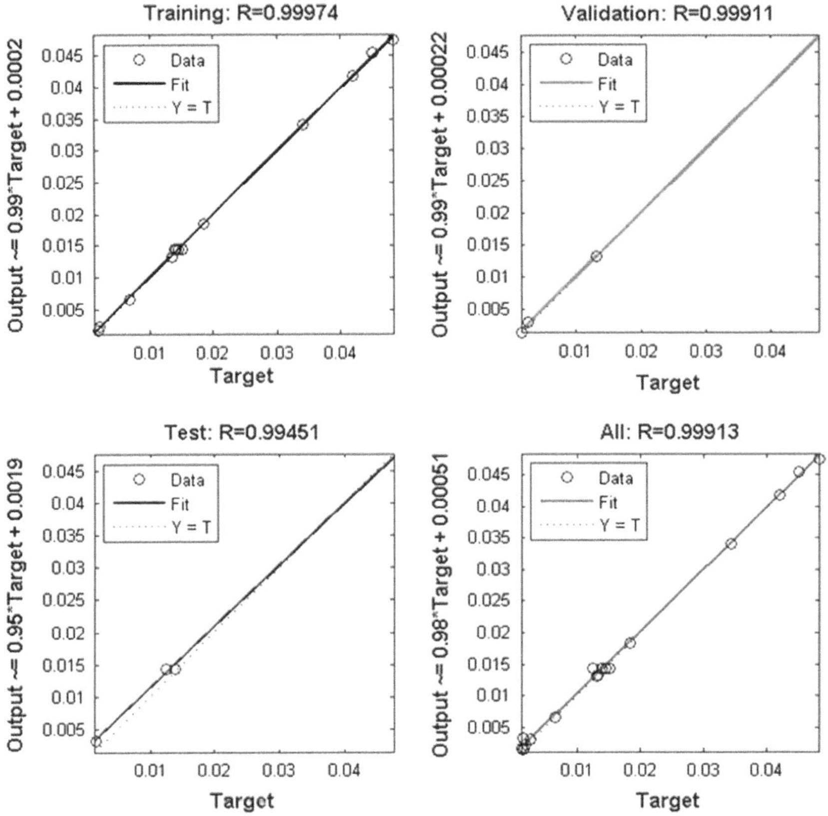


*Figure 9.5* Linear regression analysis of experimental and ANN-predicted MRR, respectively, depicts more than a 99% associated fit.

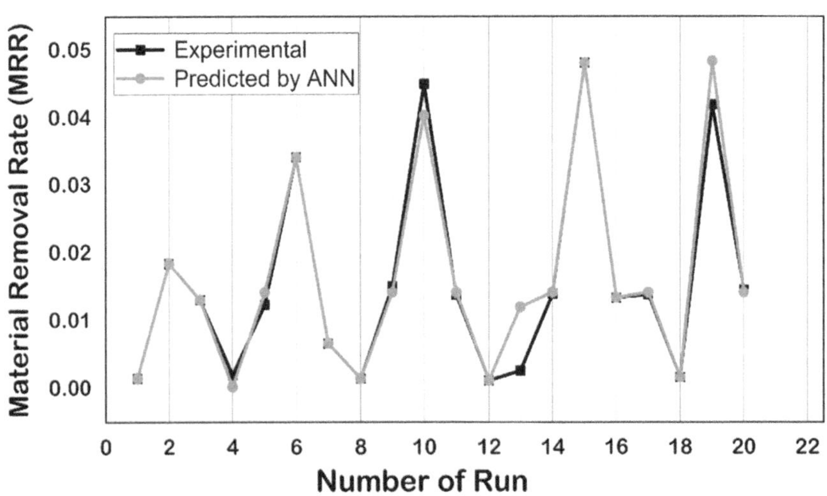


*Figure 9.6* Evaluation of predicted value relative to experimental value depicts the ANN model.

## 9.6 CONCLUSION

This research article communicates the EDM machining potential to machine AISI 202 grade stainless steel. Further optimization of machining parameters is obtained by performing response surface methodology. The machined surface and tool surface were examined by SEM and extrapolate the technicality for the improvement. Further, the ANN model was exercised to predict the experimental result and found a good agreement with the result. The optimum parameter obtained to elevate MRR was: pulse current 9 A, pulse on-time 75 μs, and pulse off-time 9 μs, respectively. Similarly, an optimum parametric setting for reducing tool wear rate was pulse current 5 A, pulse on-time 65 μs, and pulse off-time 7 μs, respectively. The SEM study revealed that at a higher current setting, MRR improved but micro-cracks were attained due to residual thermal stress. Moreover, the presence of globules and a white layer helped to reduce the erosion rate of the electrode significantly.

## REFERENCES


1. Rajmohan, T., Prabhu, R., Subbarao, G., & Palanikumar, K. (2012) Optimization of machining parameters in electrical discharge machining (EDM) of 304 stainless steel. *Procedia Engineering International Conference on Modeling, Optimization and Computing* 38: 1030–1036.
2. Muthuramalingam, T., & Mohan, B. (2014) A review on influence of electrical process parameters in EDM process. *Archives of Civil and Mechanical Engineering* 15(1): 87–94.
3. Marafona, J., & Wykes, C. (2000) A new method of optimizing material removal rate using EDM with copper–tungsten electrode. *International Journal of Machine Tool and Manufacture* 40: 153–164.
4. Ho, K. H., & Newman, S. T. (2003) State of the art electrical discharge machining (EDM). *International Journal of Machine Tool and Manufacture* 43: 1287–1300.
5. Li, L., Wong, Y. S., Fuh, J. Y. H., & Lu, L. (2001) EDM performance of TiC/copper-based sintered electrodes. *Materials and Design* 22: 669–678.
6. Khanra, A. K., Sarkar, B. R., Bhattacharya, B., Pathak, L. C., & Godkhindi, M. M. (2007) Performance of $ZrB_2$-Cu composite as an EDM electrode. *Journal of Materials Processing Technology* 183: 122–126.
7. Lopez-Esteban, S., Gutierrez-Gonzalez, C. F., Mata-Osoro, G., Pecharroman, C., Diaz, L. A., Torrecillas, R., & Moya, J. S. (2010) Electrical discharge machining of ceramics/semiconductor/metal nano composites. *Scripta Materialia* 63: 219–222.
8. Kiyak, M., & Cakir, O. (2007) Examination of machining parameters on surface roughness in EDM of tool steel. *Journal of Material Processing Technology* 191: 141–144.
9. Czelusniak, T., Amorim, F. L., Lohrengel, A., & Higa, C. F. (2014) Development and application of copper-nickel zirconium diboride as EDM

electrodes manufactured by selective laser sintering. *International Journal of Advanced Manufacturing Technology* 72: 5–8.

10. Rajurkar, K. P., Sundaram, M. M., & Malshe, A. P. (2013) Review of electrochemical and electro-discharge machining. *CIRP Conference on Electro Physical and Chemical Machining (ISEM)* 6: 13–26.
11. Fonseca, J., & Marafona, J. D. (2013) The effect of deionization time on the electric discharge machining performance. *International Journal of Advanced Manufacturing Technology* 71: 1–4.
12. Tang, L., & Guo, Y. F. (2013) Electrical discharge precision machining parameters optimization investigation on S-03 special stainless steel. *International Journal of Advanced Manufacturing Technology* 70: 5–8.
13. Tsai, H. C., Yan, B. H., & Huang, F. Y. (2003) EDM performance of Cr/Cu-based composite electrode. *International Journal of Machine Tools and Manufacture* 43: 245–252.
14. Lee, H. T., & Tai, T. Y. (2003) Relationship between EDM parameters and surface crack formation. *Journal of Material Processing Technology* 142: 676–683.
15. Zarepour, H., Tehrani, A. F., Karimi, D., & Amini, S. (2007) Statistical analysis on electrode wear in EDM of tool steel DIN 1.2714 used in forging dies. *Journal of Material Processing Technology* 187–188: 711–714.
16. Haron, C. H., GhaniJ, A., Burhanuddin, Y., Seong, Y. K., & Swee, C. Y. (2008) Copper and graphite electrode performance in electrical–discharge machining of XW42 tool steel. *Journal of Materials Processing Technology* 201: 570–573.
17. Hu, C. F., Zhou, Y. C., & Bao, Y. W. (2008) Material removal and surface damage in EDM of $Ti_3Si_2$ceramic. *Ceramics International* 34: 537–541.
18. Muttamara, A., Fukuzava, Y., Mohri, N., & Tani, T. (2009) Effect of electrode material on electric discharge machining of alumina. *Journal of Materials Processing Technology* 209: 2545–2552.
19. Bhaumik, M., & Maity, K. (2018) Effect of deep cryotreated tungsten carbide electrode and Sic powder on EDM performance of AISI304. *Particulate Science and Technology* 37: 981–992. 10.1080/02726351.2018.1487491
20. Tripathy, S., & Tripathy, D. K. (2017) Multi-response optimization of machining process parameters for powder mixed electro-discharge machining of H-11 die steel using grey relational analysis and topsis. *Machining Science and Technology* 21(3): 362–384.
21. Surekha, B., Lakshami, T. S., Jena, H., & Samal, P. (2021) Response surface modeling and application of fuzzy grey relational analysis to optimize the multi response characteristics of EN-19 machining using powder mixed EDM. *Australian Journal of Mechanical Engineering* 19(1): 19–29. 10.1080/14484846.2018.1564527
22. Kong, L., Liu, Z., Qiu, M., Wang, W., Han, Y., & Bai, S. (2019) Machining characteristics of submerged gas flushing electrical discharge machining of Ti6Al4V alloy. *Journal of Manufacturing Process* 41: 188–196.
23. Hewidy, M. S., Al-Tawel, T. A., & El-Safty, M. F. (2005) Modelling and machining parameters of wire electrical discharge machining of inconel 601 using RSM. *Journal of Materials Processing Technology* 169: 328–336.

24. Assarzadeh, S., & Ghoreishi, M. (2013) Statistical modeling and optimization of process parameters in electro-discharge machining of cobalt-bonded tungsten carbide composite. *Procedia of the 17$^{th}$ CIRP Conference on Electro Physical and Chemical Machining* 6: 463–468.
25. Samesh, S. H. (2009) Study of parameters in electric discharge machining through response surface methodology approach. *Applied Mathematical Modeling* 33: 4397–4407.
26. Arka, G. N., Sahoo, S. K., Iqbal, M. M., & Singh, S. (2021) Acoustic horn tool assembly design for ultrasonic assisted turning and its effects on performance potential. *Materials and Manufacturing Processes* 37(3): 260–270. DOI:10.1080/10426914.2021.2016819
27. Ekmkci, B. (2007) Residual stress and white layer in electric discharge machining (EDM). *Applied Surface Science* 253: 9234–9240.
28. Mannan, K. T., Krishnaiah, A., & Arikatla, S. P. (2013) Surface characterization of electric discharge machined surface of high speed steel. *Advanced Materials Manufacturing and Characterization* 3.
29. Bharti, P. S. (2019) Process modelling of electric discharge machining by back propagation and radial basis function neural network. *Journal of Information & Optimization Science* 40(2): 263–278.
30. Singh, B., & Mishra, J. P. (2019) Surface finish analysis of wire electric discharge machined specimens by RSM and ANN modeling. *Measurement* 137: 225–237.

Chapter 10

# Prediction and Neural Modeling of Material Removal Rate in Electrochemical Machining of Nimonic-263 Alloy

*Dilkush Bairwa*[1], *Dr Ravi Pratap Singh*[1], *Dr Ravinder Kataria*[2], *Dr Ravi Butola*[3], *Dr Mohd Javaid*[4], *Shailendra Chauhan*[1], *and Madhusudan Painuly*[1]

[1]Department of Industrial and Production Engineering, Dr B R Ambedkar National Institute of Technology, Jalandhar, Punjab, India
[2]National Institute of Fashion Technology, Srinagar, Jammu & Kashmir, India
[3]Department of Mechanical Engineering, Delhi Technological University, Delhi, India
[4]Department of Mechanical Engineering, Jamia Millia Islamia, New Delhi, India

**CONTENTS**



## 10.1 INTRODUCTION

Because there is contactless operating, micro-electrochemical machining (micro-ECM) may be able to produce the greater surface integrity. Numerous physical and chemical processes, including heat transfer, electrolyte flow, electrochemical reactions, and electric field transmission, among others, take place concurrently during micro-electrochemical machining. Faraday's law of electrolysis serves as the foundation for the ECM process. During the

DOI: 10.1201/9781003258421-11

electrochemical machining operation ($NaNO_3$), the tool electrode and the workpiece are both submerged in an electrolytic concentration of sodium nitrate [1]. We have managed a tiny gap between the two electrodes with no tool wear, mechanical forces, or residual stresses by applying voltage at a consistent value between the tool electrode and the workpiece. In electrochemical machining (ECM), the workpiece acts as an anode (positive) and the tool electrode as a cathode side. When DC power is applied between the two, the anodic material may be dissolved.

### 10.1.1 Micro-Electrochemical Machining

Three-dimensional microstructures like micro-holes and micro-grooves may be created using a micro-electrochemical machining method. Many industrial uses, particularly aerospace engineering, automotive engineering, and microelectronics, use micro-electrochemical machining. Figure 10.1

The workpiece is set in a frame, the tool is introduced into a tool post, and both are held extremely near to one another so that the inter-electrode spacing may be maintained. The inter-electrode gap (IEG) is the space between the tool and the workpiece. A potential difference is introduced around across tool and workpiece after submerging them in a liquid electrolyte (such as NaCl), and material removal starts.

### 10.1.2 Materials

Due to its strong corrosion resistance, high tensile strength, high temperature resistance, and high yield strength, Nimonic-263 alloy was chosen for this experiment. Casing, rings, and fabrication sheets made of the Nomonic-263 alloy have been utilised in the aerospace applications and on aviation turbine engines. The alloy Nimonic-263 has strong heat conductivity. It is ductile at high temperatures in welded constructions.

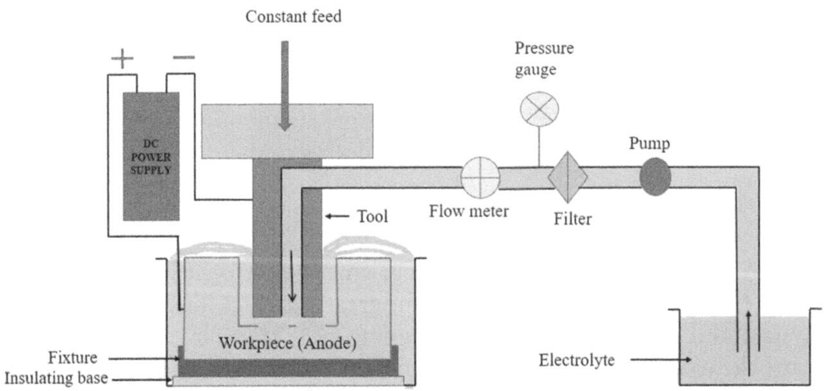


*Figure 10.1* Schematic diagram of ECM.

The surface quality has been improved with feed rate of 0.4 mm/min. Thomas Bergs et al. 2020 [2] Conducted ECM is working on pure cementite material (Fe3c) with tool material of 42CrMo4 which is kept in rotation with the speed of 20 to 40 m/s and in this study, two major machining parameters (such as feed rate $V_f$ and voltage U) have been studied and the conclusion of a process certificate for electrochemical machining has been effectively carried out for the change of the localized phase concentration and surface grinding. Shunda Zhan et al. 2020 [3] Utilizing certain machining parameters, like inlet/outlet pressure (0.30/0.08 MPa), applied voltage (20 V), feed rate (0.3 mm/min), and initial gap, the performed ECM process enhanced the machining accuracy of the blade leading edge by 41.79% (0.5 mm). Wenjian Cao and others 2020 [4] Carried counter-rotating electrochemical machining to create thin-walled revolving parts. The beginning inter-electrode gap decreased (0.272–0.158 mm) with respect to increasing (11–15 V), and the inter-electrode gap continued to increase after the transitional period because the relative equilibrium MRR is greater than the corresponding feed rate. X.L. Chen and others 2020 [5,6] Three electrolyte flow modes and induced ECM of microchannels with coated porous cathodes have been used, with jet flow type pulse duty cycle 20% more beneficial to reducing the surface roughness of the microgrooves, standard error, as well as machining parameters such as temperature (250 C), frequency (2 kHz), duty cycle (20%, 40%, 60%, 80%), machining time (10 s), voltage (10, 15, 20, 25 V), and pressure (0.4 MPa). Wang, Feng, et al. 2020 [7] Utilising synchronised electrochemical machining, a spheroid hole's surface quality could be enhanced, and by raising, the amplitude of the sidewall and the curvature of the fillet were essentially lowered, affecting the efficiency of the rhombus hole.

## 10.2 MATERIALS AND METHODS

To carry out the experimental investigation on the Nimonic-263 alloy with micro-ECM. A workpiece made of the Nimonic-263 alloy measuring 1 cm × 1 cm has been utilised for the experiments. Micro-drilling operations have been carried out using micro-electrochemical machining. The various input process variables, such as applied voltage, feed rate, pulse frequency, and electrolyte solution, have been taken into consideration to conduct an experiment to examine the effects of machining inputs in order to obtain micro-holes with a high precision profile and acceptable surface quality. The material's elemental makeup is listed in Table 10.1. Nimonic-263, a nickel-based superalloy, has been subjected to micro-electrochemical machining for testing. On the basis of the Taguchi L16 orthogonal array method [8], the experiments were created. We chose four parameters with four fixed levels. Initial holes were drilled into the Nimonic-263 alloy that use the micro-ECM method and the optimum settings, and each hole was subsequently machined

*Table 10.1* Chemical composition of Nimonic-263 alloy

| *Sr. no.* | *Elements* | *Content (%)* |
|---|---|---|
| 1 | Nickel, Ni | 49 |
| 2 | Cobalt, Co | 19–21 |
| 3 | Chromium, Cr | 19–21 |
| 4 | Titanium, Ti | 1.90–2.40 |
| 5 | Iron, Fe | 0.70 |
| 6 | Manganese, Mn | 0.60 |
| 7 | Silicon, Si | 0.40 |
| 8 | Copper, Cu | 0.20 |
| 9 | Sulphur, S | 0.0070 |

independently. After multiple tries, the ideal parameters for micro-drilling have been achieved. In this experiment, sodium nitrate ($NaNO_3$) was employed as the electrolyte solution and copper has been used as the tool electrode. This little ECM system has a maximum voltage of 10 volts, a pulse frequency range of 5 to 150 kilohertz, and a feed rate range of 20 to 50 microns per minute. For further investigation, the output response's average value has been selected. The material removal rate (MRR) has been investigated as a key output response variable in this experimental investigation. In this experiment, the impacts of process input parameters on response parameters, such as material removal rate, have been examined using Taguchi's L16 orthogonal array approach (MRR).

The inputs used were the applied voltage, pulse frequency, electrolyte concentration, and feed rate [8]. Factors are categorised on four different levels. For several experiments, the electrolyte solution has been regularly replaced with brand-new electrolyte solution. Table 10.2 displays the chosen machine inputs together with the identified level of each input. The proposed L16 orthogonal array (OA) for carrying out the tests is shown in Table 10.3. To compare productivity and accuracy, the micro-holes were made using a systematic technique, and the material removal rate was used to gauge the significance of the results.

*Table 10.2* Machining input parameters with selected levels of micro-ECM

| *Parameters* | *Level 1* | *Level 2* | *Level 3* | *Level 4* |
|---|---|---|---|---|
| Feed rate (mm/min) | 0.02 | 0.03 | 0.04 | 0.05 |
| Applied voltage (V) | 7 | 8 | 9 | 10 |
| Pulse frequency (kHz) | 50 | 75 | 100 | 125 |
| Electrolyte concentration (g/l) | 40 | 50 | 60 | 70 |

*Table 10.3* L16 orthogonal array of micro-ECM with average output response (avg. MRR)

| *Exp. no.* | *AV (volt)* | *EC (g/l)* | *PF (kHz)* | *FR (mm/min)* | *Average MRR ($mm^3$/min)* |
|---|---|---|---|---|---|
| 1. | 7 | 40 | 50 | 0.02 | 0.070500 |
| 2. | 7 | 50 | 75 | 0.03 | 0.112533 |
| 3. | 7 | 60 | 100 | 0.04 | 0.153723 |
| 4. | 7 | 70 | 125 | 0.05 | 0.197400 |
| 5. | 8 | 40 | 75 | 0.04 | 0.126093 |
| 6. | 8 | 50 | 50 | 0.05 | 0.188967 |
| 7. | 8 | 60 | 125 | 0.02 | 0.113733 |
| 8. | 8 | 70 | 100 | 0.03 | 0.123733 |
| 9. | 9 | 40 | 100 | 0.05 | 0.127400 |
| 10. | 9 | 50 | 125 | 0.04 | 0.152633 |
| 11. | 9 | 60 | 50 | 0.03 | 0.153500 |
| 12. | 9 | 70 | 75 | 0.02 | 0.133567 |
| 13. | 10 | 40 | 125 | 0.03 | 0.137233 |
| 14. | 10 | 50 | 100 | 0.02 | 0.142267 |
| 15. | 10 | 60 | 75 | 0.05 | 0.159933 |
| 16. | 10 | 70 | 50 | 0.04 | 0.188167 |

## 10.3 RESULTS AND DISCUSSION

The signal to noise (S/N) ratio has been used to calculate the effect of parameter level on output parameter. Micro-ECM setup via a specific set of 16 experiments may be used to drill micro-holes.

Tables 10.3 and 10.4 both show the average material removal rate and the S/N ratio for every output response, respectively. The highest mean S/N ratio among the values determines the factor's optimal level [8]. The MRR for SN ratios ANOVA findings are shown in Table 10.4. The material removal rate (MRR) for means and the ANOVA findings are shown in Table 10.5.

*Table 10.4* ANOVA results for MRR (SN ratios)

| *Source* | *DF* | *Seq. SS* | *Adj. SS* | *Adj. MS* | *F* | *P* |
|---|---|---|---|---|---|---|
| AV (volt) | 3 | 7.896 | 7.896 | 2.6321 | 0.73 | 0.601 |
| EC (g/l) | 3 | 19.736 | 19.736 | 6.5787 | 1.81 | 0.319 |
| PF (kHz) | 3 | 2.031 | 2.031 | 0.6770 | 0.19 | 0.899 |
| FR (mm/min) | 3 | 29.292 | 29.292 | 9.7639 | 2.69 | 0.219 |
| Residual error | 3 | 10.883 | 10.883 | 3.6278 | | |
| Total | 15 | 69.838 | | | | |

*Table 10.5* ANOVA results for MRR (means)

| *Source* | *DF* | *Seq. SS* | *Adj. SS* | *Adj. MS* | *F* | *P* |
|---|---|---|---|---|---|---|
| AV (volt) | 3 | 0.001229 | 0.001229 | 0.000410 | 0.54 | 0.687 |
| EC (g/l) | 3 | 0.004489 | 0.004489 | 0.001496 | 1.97 | 0.296 |
| PF (kHz) | 3 | 0.000972 | 0.000972 | 0.000324 | 0.43 | 0.749 |
| FR (mm/min) | 3 | 0.006812 | 0.006812 | 0.002271 | 2.99 | 0.196 |
| Residual error | 3 | 0.002276 | 0.002276 | 0.000759 | | |
| Total | 15 | 0.015779 | | | | |

### 10.3.1 Material Removal Rate for Means

This answer establishes the process performance. A high MRR denotes greater output, and Figure 10.2's major effect chart shows: It has been determined that the following input parameters are the best ones: applied voltage (10 V), electrolyte concentration (70 g/l), pulse frequency (50 kHz), and feed rate (0.05 mm/min). The input machining parameters A4, B4, C1, and D4 indicate. With these settings, the material removal process may be enhanced. Less current would be created at low voltage, which would result in less material being removed. A higher signal-to-noise ratio is preferable for MRR optimisation. Figure 10.2

This graph displays the means' major impacts plot while also noting the MRR values:

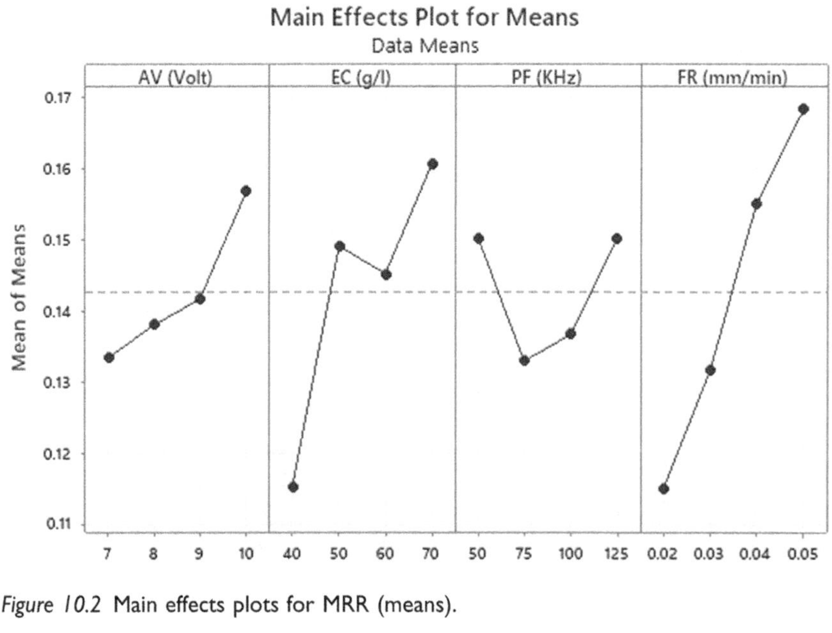


*Figure 10.2* Main effects plots for MRR (means).

- The quantity of material removed continuously increased as the voltage value raised from 7 to 9 volts, but beyond that point, the MRR values only slightly increased. Minimum MRR at level one's 7 V.
- When the electrolyte concentration is raised from 40 g/l to 50 g/l, the quantity of material will rise. As a result, the MRR will slightly increase until the electrolyte concentration is increased to 60 g/l.
- The most quantity of material may be eliminated at the lowest pulse frequency frequency of 50 kHz, following which the MRR will slightly reduce up to the pulse frequency of 125 kHz before increasing once more at pulse frequencies of 75 kHz and 125 kHz.
- This graph shows how feed rate affects MRR over time as MRR increases continuously up to a feed rate of 0.05 mm/min.

### 10.3.2 Material Removal Rate for SN Ratios

This graph showing the main effects plot for SN ratios and it is observing the MRR values: Figure 10.3

### 10.3.3 Residual Plots for Means

This graph displays the mean ratio residual plots and analyses the MRR values across several parameters.

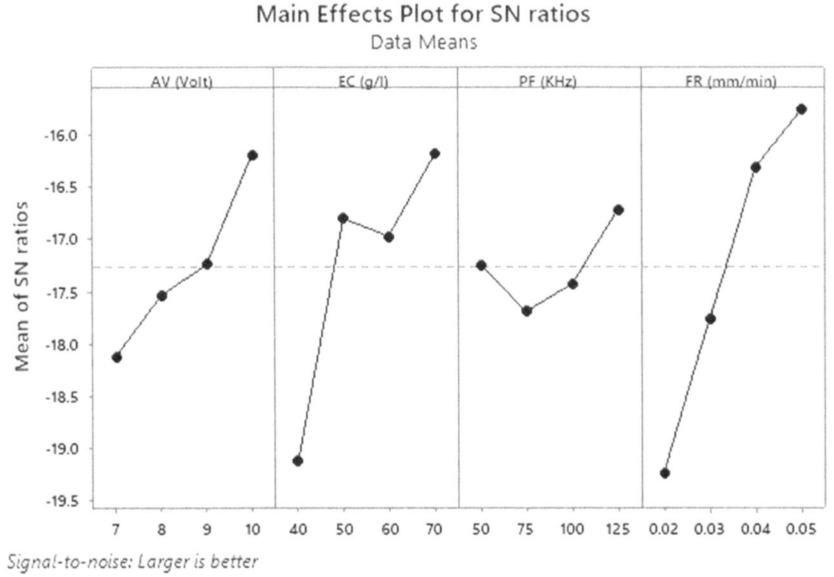


*Figure 10.3* Main effects plots for MRR (SN ratios).

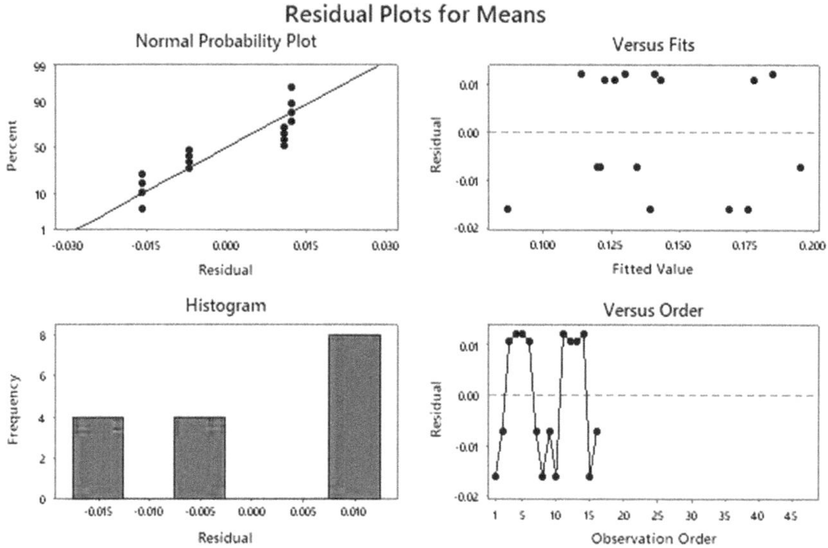


*Figure 10.4* Residual plots for MRR (means).

- Figure 10.4 displays the residuals' normal probability charts for MRR. The normally distributed curve is shown as a straight line. MRR values between the normal percentage of probability and the residuals are illustrated by the normal probability plot graph. The normal probability plot, which depicts the polynomial regression model and keeps all of the data points in close proximity to the polynomial regression line, is used to analyse the experimental MRR values.
- The MRR values between the residuals and fitted values graph are shown by the versus fits graph. During a multiple regression, a "residuals versus fits plot" is the most typical visual produced. This figure may be used to find non-linearity, mismatched error variances, and outliers. The fitted values are on the x-axis, while the residuals are on the y-axis.
- The histogram displays a frequency versus residuals graph to depict the MRR values. The residual histogram may be used to examine how evenly the variation is spread.
- The connection between the residual and the observation order is shown by the versus order plot. In order to detect the mistake in the sequence that is near to each other, a residual versus observation order graph was created regardless of how the data were gathered in space.

### 10.3.4 Residual Plots for SN Ratios

The MRR values are being observed at various factors while this graph displays the residual plots for SN ratios. Similar to the residual plot for means, the residual plot for SN ratios represents all values. Figure 10.5

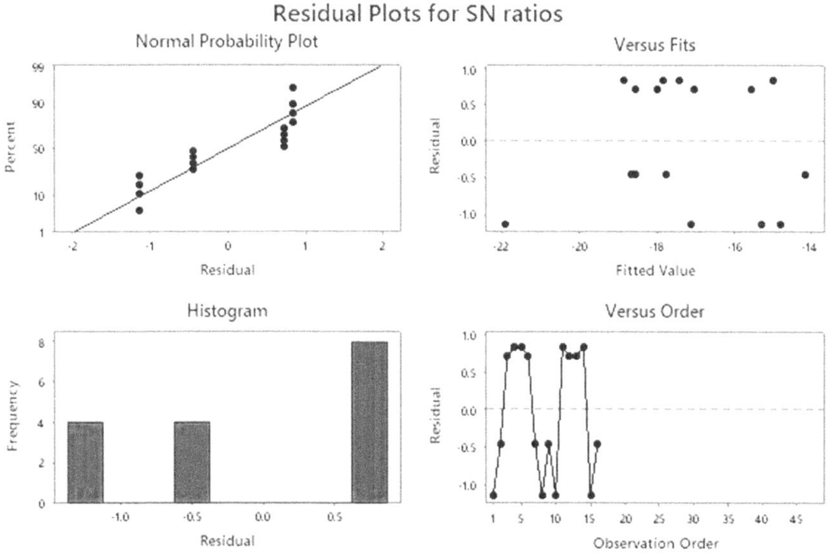


*Figure 10.5* Residual plots for MRR (SN ratios).

## 10.4 PREDICTION OF MRR VALUES THROUGH THE ANN MODEL

Based on the input parameters such as feed rate, electrolyte concentration, pulse frequency, and applied voltage. The material removal rate has been calculated at the time of machining of Nimonic-263 alloy. A multi-layer neural network is where two hidden layers were selected of 20 neurons each is added to predict the data. The ANN model has been trained by the Bayesian regulation of MATLAB. This regularisation can handle improper data by solving the problems underfitting and overfitting. Table 10.6 shows the statistical data of MRR for ANN.

### 10.4.1 Artificial Neural Network Model

The neural network used as shown in Figure 10.6 has four input neurons such as feed rate, electrolyte concentration, pulse frequency, and applied voltage also has two hidden layers containing 20 neurons each (as per accuracy level) and finally, there is one output neuron (MRR). For training the data, the Levenberg-Marquardt algorithm was used for better results, and further random data division and for performance mean square error are used as parameters for ANN. According to the Figure 10.6, w and b are the assigned weights and biases for the neurons; further, the value of the material removal rate has been predicted by using the 4–20-20-1 structure

*Table 10.6* Statistical data of MRR for ANN

| *Exp. no.* | *AV (volt)* | *EC (g/l)* | *PF (kHz)* | *FR (mm/min)* | *AVERAGE MRR ($mm^3/min$)* | *ANN MRR* |
|---|---|---|---|---|---|---|
| 1 | 7 | 40 | 50 | 0.02 | 0.0705 | 0.15624 |
| 2 | 7 | 50 | 75 | 0.03 | 0.112533 | 0.11253 |
| 3 | 7 | 60 | 100 | 0.04 | 0.153723 | 0.15372 |
| 4 | 7 | 70 | 125 | 0.05 | 0.1974 | 0.16817 |
| 5 | 8 | 40 | 75 | 0.04 | 0.126093 | 0.12609 |
| 6 | 8 | 50 | 50 | 0.05 | 0.188967 | 0.18897 |
| 7 | 8 | 60 | 125 | 0.02 | 0.113733 | 0.11373 |
| 8 | 8 | 70 | 100 | 0.03 | 0.123733 | 0.12373 |
| 9 | 9 | 40 | 100 | 0.05 | 0.1274 | 0.1274 |
| 10 | 9 | 50 | 125 | 0.04 | 0.152633 | 0.15263 |
| 11 | 9 | 60 | 50 | 0.03 | 0.1535 | 0.1535 |
| 12 | 9 . | 70 | 75 | 0.02 | 0.133567 | 0.13357 |
| 13 | 10 | 40 | 125 | 0.03 | 0.137233 | 0.16098 |
| 14 | 10 | 50 | 100 | 0.02 | 0.142267 | 0.12612 |
| 15 | 10 | 60 | 75 | 0.05 | 0.159933 | 0.15993 |
| 16 | 10 | 70 | 50 | 0.04 | 0.188167 | 0.18817 |

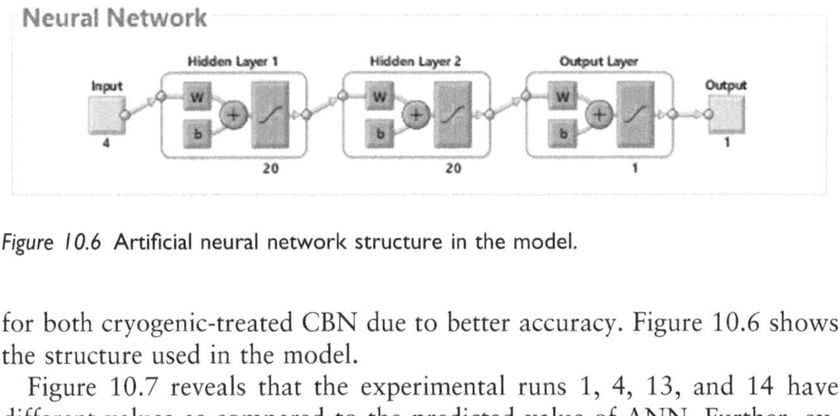


*Figure 10.6* Artificial neural network structure in the model.

for both cryogenic-treated CBN due to better accuracy. Figure 10.6 shows the structure used in the model.

Figure 10.7 reveals that the experimental runs 1, 4, 13, and 14 have different values as compared to the predicted value of ANN. Further, experiment numbers 1, 4, 6, 13, and 16 have higher MRR values. Figure 10.7 shows the comparison between predicted values through ANN and experimental values for NC-CBN.

Figure 10.8 shows the plot between mean square error (MSE) and six epochs with a non-cryogenic-treated CBN data set. In ANN, an epoch is one cycle through the full training data set. Figure 10.8 further shows that the best validation has occurred at one epoch, which means one cycle is implemented for full training of the data set in the case of non-cryogenic-treated CBN.

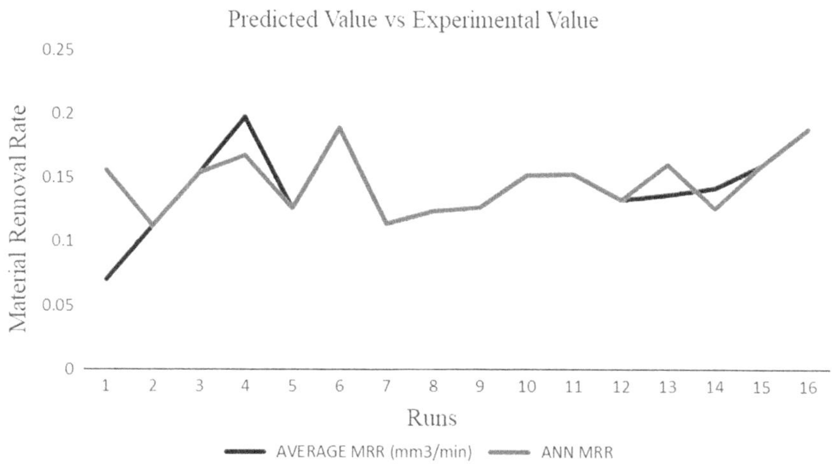


*Figure 10.7* Comparison of predicted and experimental values of MRR.

Figure 10.8 reveals the case of non-cryogenic-treated CBN when predicted by the ANN model as the experimental values.

Figure 10.9 shows the comparison between predicted values through ANN and experimental values for non-cryogenic CBN. Figure 10.9 shows the plot between target value and output value (predicted values).

## 10.5 CONCLUSION

The current experimental work presents the experimental technique of machining input parameters to evaluate the material removal rate. Micro-drilling operations have been carried out to produce micro-holes on Nimonic-263 alloy. A systematic strategy was employed to collect all of the data needed for experimentation and design. An artificial neural network model can be used to determine the predicted data of material removal rate and can also be used to evaluate the error between experimental values and predicted values of material removal rate. The findings are based on experimental data at four levels and the Taguchi L16 orthogonal technique to determine the best process parameters.

- The optimal parametric observation for meanings A4, B4, C1, and D4 (applied voltage 10 V, electrolyte concentration 70 g/l, pulse frequency 50 kHz, and feed rate 0.05 mm/min) has been explored based on the micro-ECM operation.
- The optimal parametric observation has been examined for SN ratios A4, B4, C4, and D4 based on the micro-ECM operation.

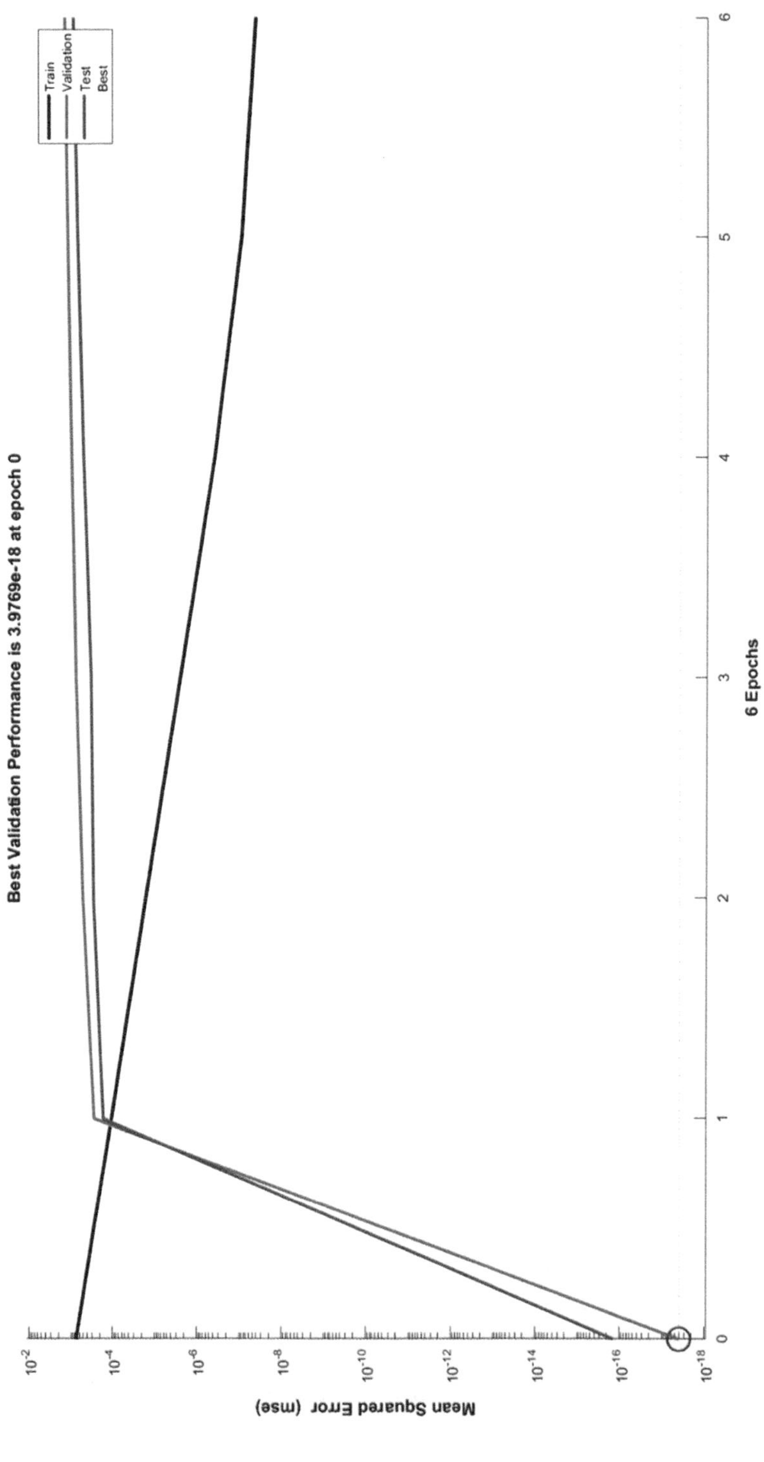


*Figure 10.8* Iteration graph for MRR through ANN.

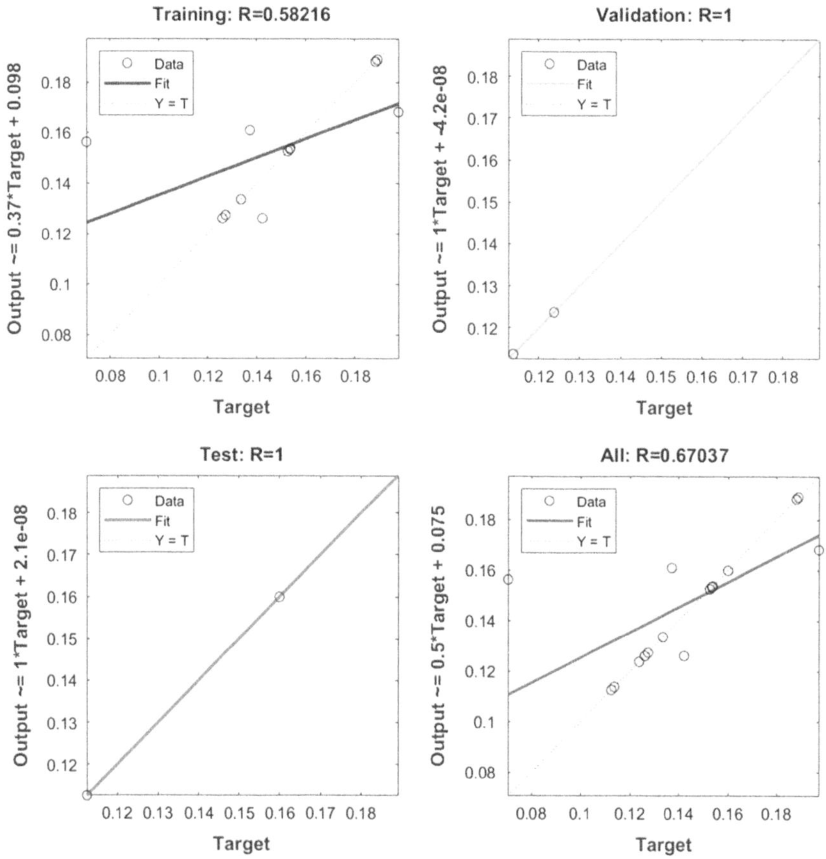


*Figure 10.9* Regression plots through ANN of NC-CB for MRR.

- The two factors that have the most effects on the workpiece are the supplied voltage and feed rate.
- In the current experiment, the most material may be removed when the following conditions are met: applied voltage of 10 V, electrolyte concentration of 70 g/l, pulse frequency of 50 kHz, and feed rate of 0.05 mm/min.
- By employing MINITAB software, better outcomes in terms of design of experiments (DOE) methodology have been obtained.
- In order to compute predicted MRR values, an artificial neural network (ANN) model was used.
- Due to its higher accuracy, the 4–20-20-1 structure has been used to forecast the value of the material removal rate for both cryogenically treated CBN.

## REFERENCES

1. Y. Wang, Z. Xu, and A. Zhang, "Anodic characteristics and electrochemical machining of two typical γ-TiAl alloys and its quantitative dissolution model in NaNO3 solution," *Electrochim. Acta*, vol. 331, p. 135429, 2020, doi: 10.1016/j.electacta.2019.135429
2. T. Bergs and S. Harst, "Development of a process signature for electrochemical machining," *CIRP Ann.*, vol. 69, no. 1, pp. 153–156, 2020, doi: 10.1016/j.cirp.2020.04.078
3. S. Zhan and Y. Zhao, "Plasma-assisted electrochemical machining of microtools and microstructures," *Int. J. Mach. Tools Manuf.*, vol. 156, no. 1088, p. 103596, 2020, doi: 10.1016/j.ijmachtools.2020.103596.
4. W. Cao, D. Wang, and D. Zhu, "Modeling and experimental validation of interelectrode gap in counter-rotating electrochemical machining," *Int. J. Mech. Sci.*, vol. 187, no. June, p. 105920, 2020, doi: 10.1016/j.ijmecsci.2020.105920
5. X. L. Chen *et al.*, "Investigation on the electrochemical machining of micro groove using masked porous cathode," *J. Mater. Process. Technol.*, vol. 276, no. May 2019, 2020, doi: 10.1016/j.jmatprotec.2019.116406
6. G. Liu, H. Tong, Y. Li, and H. Zhong, "Novel structure of a sidewall-insulated hollow electrode for micro electrochemical machining," *Precis. Eng.*, vol. 72, no. April, pp. 356–369, 2021, doi: 10.1016/j.precisioneng.2021.05.009
7. F. Wang, J. Yao, and M. Kang, "Electrochemical machining of a rhombus hole with synchronization of pulse current and low-frequency oscillations," *J. Manuf. Process.*, vol. 57, no. 40, pp. 91–104, 2020, doi: 10.1016/j.jmapro.2020.06.014
8. P. V. Sivaprasad, K. Panneerselvam, and A. Noorul Haq, "A comparative assessment in sequential μ-drilling of Hastelloy-X using laser in combination with μ-EDM and μ-ECM," *J. Brazilian Soc. Mech. Sci. Eng.*, vol. 43, no. 7, pp. 1–19, 2021, doi: 10.1007/s40430-021-03068-4

Chapter 11

# Optimization of End Milling Process Variables Using a Multi-Objective Genetic Algorithm

*Jignesh Girishbhai Parmar*[1] *and*
*Dr. Komal Ghanshyambhai Dave*[2]
[1]Gujarat Technological University, Ahmedabad, Ahmedabad, Gujarat, India
[2]L D Engineering College, Ahmedabad, Gujarat, India

**CONTENTS**



## 11.1 INTRODUCTION

The various simple to complex components have been machined on a milling machine with various types of machining operations. End milling operation is used to produce slots, flat surfaces and profiles by using an end milling cutter. Reddy Sreenivasulu et al. [2019] [1] used the grey-based Taguchi approach integrated with entropy measurement for glass fiber–reinforced polymer matrix composite for optimization. S. Ajith Arul Daniel et al. [2019] [2] used ANN and the Taguchi grey relation analysis to optimize the input variables like cutting speed, feed, depth of cut, mass fraction, and particle size of SiC for the milling process of Al5059-SiC-MoS2. I.A. Daniyan et al. [2021] [3] used response surface methodology to find out the correlation of the end milling process parameters in terms of optimization. The experiment has been conducted on the DMU80monoBLOCK Deckel Maho 5-axis CNC milling machine for AA6063. The mathematical and optimization model has been developed based on numerical value and physical experiment results. Thanongsak Thepsonthi et al. [2021] [4] investigated micro end milling operation for titanium-based alloy. Statistical model and particle swarm

DOI: 10.1201/9781003258421-12 

optimization have been used for achieving the optimum condition for minimum surface roughness with minimum burr formation. Ti-6Al-4V titanium alloy has been used for experimentation. Trung-Thanh Nguyen et al. [2020] [5] used adaptive simulated annealing for optimization of end milling process parameters to improve the energy efficiency and power factor with minimum surface roughness. Radius basic neural network has been used for prediction purposes. Rajeswari S. et al. [2018] [6] used grey fuzzy logic for multi-objective optimization of cutting variable for minimum surface roughness and tool wear. Jakeer Hussain Shaik et al. [2018] [7] used a finite element model to improve the dynamic stability of end milling operation. Analysis of variance and artificial neural network have been used to study the effect of input variables on output and prediction. Nuraini Lusi et al. [2020] [8] optimized end milling process parameters for ASSAB-XW 42. The Taguchi method with grey relational analysis has been used for multi-response optimization. M. Jebaraj et al. [2019] [9] investigated the effect of input variables on responses for end milling operation on Al 6082-T6 alloy. They used TOPSIS and analysis of variance for statistical analysis. It reveals that some work has been done for prediction and optimization of the milling process with experimentation. It is also found that, after the experimentation, researchers have used different artificial intelligence techniques for specified materials. Additionally, the use of input parameters of the machining process with mechanical property of material are not involved in previous procedures.

## 11.2 MATHEMATICAL MODEL

Mathematical model has been developed based on input variables of the end milling process. The tool life has been calculated through Equation (11.1) [10].

$$\textit{Cutting Speed} \times \text{Tool } \textit{life}^{\text{n}} = \text{c} \tag{11.1}$$

The MRR has been calculated through Equation (11.2) [11].

$$\text{MRR} = \frac{\textit{Initial weight of the work piece} - \textit{final weight of the work piece}}{\textit{Material density} \times \textit{Machining time}} \tag{11.2}$$

Tangential cutting force has been calculated through Equation (11.3) [12].

$$F = \begin{pmatrix} \textit{Number of teeth with simul}\ \tan\ \textit{eous engagement} \\ \textit{with the workpiece} \times \textit{Specific cutting force} \times \\ \textit{Average chip thickness} \times \textit{Chip width} \end{pmatrix} \tag{11.3}$$

The torque has been calculated through Equation (11.4) [12].

$$T_q = \frac{Tangential\ cutting\ force \times Cutter\ diameter}{2} \quad (11.4)$$

Power (Kw) has been calculated through Equation (11.5) [12].

$$\mathrm{P} = \frac{Tangential\ cutting\ force \times Cutting\ Speed}{6120} \quad (11.5)$$

The machining time has been calculated through Equation (11.6) [13].

$$\mathrm{T} = \left( \frac{Work\ piece\ length + Approach\ distance + Over\ run + Cutter\ diameter/2}{Feed\ rate \times Total\ number\ of\ teeth\ of\ end\ mill \times Spindle\ speed} \right) \times 60 \quad (11.6)$$

Where n and c are constants determined by the work tool pair and cutting condition.

The DOE is a logical methodology to determine the arrangement of machining variables for required outputs for investigation aspects [14]. Here, the full factorial method has been implemented for the DOE. For a full factorial design, if the L is the numbers of levels and M is the number of factors, then the possible design N is

$$N = L^M \quad (11.7)$$

Here five numbers of levels and three numbers of factors have been selected and based on Equation (11.7) for a total of 125 combinations have been achieved through Minitab 17 software. A design data book has been used for a range selection of cutting variables [15]. In this study, selected factors are cutting speed with five levels (140, 150, 160, 170, 180), feed with five levels (0.12, 0.15, 0.18, 0.21, 0.24), depth of cut with five levels (0.2, 0.4, 0.6, 0.8, 1), and also AISI1020 material properties such as density and hardness have been selected.

## 11.3 MULTI-OJECTIVE OPTIMIZATION

In industries required the optimization of multiple objectives instead of a single objective. MATLAB R2015a with the MOGA tool has been utilized for the optimization of multi-responses. The objective functions have been generated for required maximum and minimum values of responses with its constraints. A multi-objective empirical-based mathematical model has been achieved based on a mathematical model.

### 11.3.1 To Maximize

$$T_l = \left[\frac{n}{a}\right]^{(1/C)} \tag{11.8}$$

$$\text{MRR} = \frac{(f1 - f2)}{\left[\text{e} \times \left(\frac{\text{l} + \text{h}/2 + \text{h}/2 + \text{h}/2}{\text{b} \times \text{j} \times \text{g}}\right) \times 60\right]} \tag{11.9}$$

### 11.3.2 To Minimize

$$\text{F} = \left[\frac{(\text{k} \times d \times \text{c} \times 57.3 \times \text{b} \times 2 \times 9.80665)}{60}\right] \tag{11.10}$$

$$T_q = \left[\left(\frac{(\text{k} \times d \times \text{c} \times 57.3 \times \text{b} \times 2 \times 9.80665)}{60}\right) \times \left(\frac{h}{2}\right)\right] \tag{11.11}$$

$$\text{P} = \left[\left(\frac{(\text{k} \times d \times \text{c} \times 57.3 \times \text{b} \times 2 \times 9.80665)}{60}\right) \times \left(\frac{a}{6120}\right)\right] \tag{11.12}$$

$$\text{T} = \left(\frac{\text{i} + \text{h}/2 + \text{h}/2 + \text{h}/2}{\text{b} \times \text{j} \times \text{g}}\right) \times 60 \tag{11.13}$$

The constraint function of responses is shown in Equation (11.14).

$$\begin{aligned} C = [&25 - ((f1 - f2)/(e \times ((i + (h/2) + (h/2) + (h/2))/(j \times b \times n)) \times 60)); \\ &((i + (h/2) + (h/2) + (h/2))/(j \times b \times n)) \times 60 - 3.40; \\ &(15 - (n/a)^{(1/C)}); \\ &((k \times d \times c \times 57.3 \times b \times 2 \times 9.80665)/60) - 364; \\ &(((k \times d \times c \times 57.3 \times b \times 2)/60) \times (h/2)) - 210; \\ &(((k \times d \times c \times 57.3 \times b \times 2)/60) \times a)/6120 - 0.95] \end{aligned} \tag{11.14}$$

The selected lower and upper bounds of 14 process variables have been shown in Table 11.1

Multi-objective genetic algorithm operator and parameters with its function are indicated in Table 11.2.

*Table 11.1* Upper and lower bound of process variable of the end milling process

| *Process variable* | *Lower bound* | *Upper bound* |
|---|---|---|
| Cutting speed (a) | 140 | 180 |
| Feed (b) | 0.12 | 0.24 |
| Depth of cut (c) | 0.2 | 1 |
| Specific cutting force (d) | 120 | 120 |
| Density (e) | 0.0079 | 0.0079 |
| Initial weight (f1) | 376 | 376 |
| Final Weight (f2) | 372.8560 | 375.3712 |
| Spindle speed (g) | 3715.4989 | 4777.0701 |
| Cutter Dia. (h) | 12 | 12 |
| Work piece length (i) | 100 | 100 |
| Total no. of teeth (j) | 4 | 4 |
| Constant n | 292 | 292 |
| Constant C | 0.18 | 0.18 |
| No. of teeth with simultaneous engaged with workpiece (k) | 0.6667 | 0.6667 |

*Table 11.2* MOGA operational parameters

| *Operational parameters* | *Values* |
|---|---|
| Creation function | Feasible population |
| Population | 50 |
| Function tolerance | 1e-4 |
| Constraint tolerance | 1e-3 |
| Cross over fraction | 0.8 |
| Migration fraction | 0.2 |
| Selection function | Tournament |
| Tournament size | 2 |
| Crossover fun. | Single point |
| Mutation fun. | Adaptive feasible |

Here, Figure 11.1 indicates the average distance between individuals. In this graph, all of the distance between the individuals is shown between 0 to 600.

Figure 11.2 gives the information about the individuals and their distance. Here, distance is 0 to 0.8 and individuals 0 to 50 show the relation between the distances of the individual's value. Figure 11.3 shows the average spread is 0.0284884 and has the maximum spread and the generations are 114.

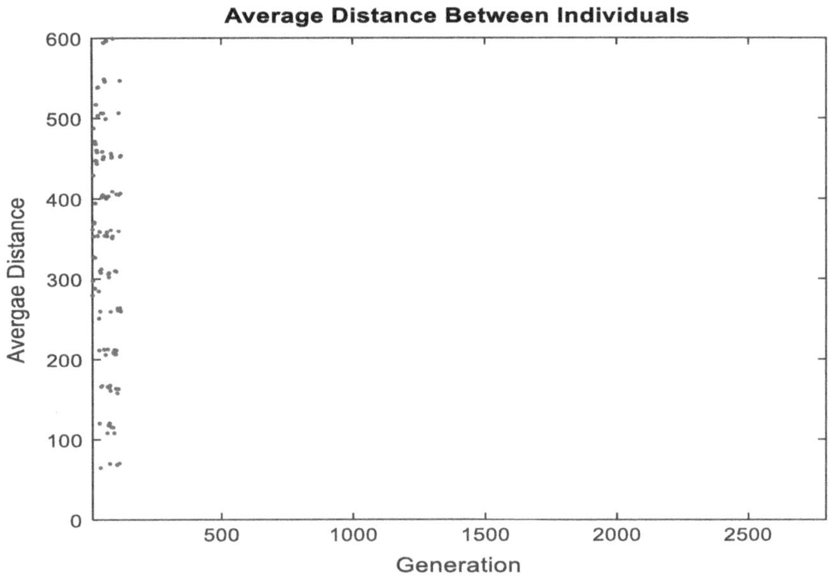


*Figure 11.1* Average distance between individuals.

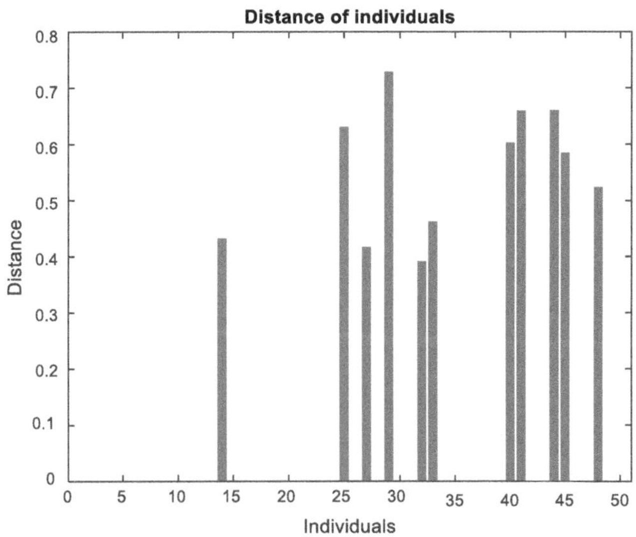


*Figure 11.2* Distance of individuals.

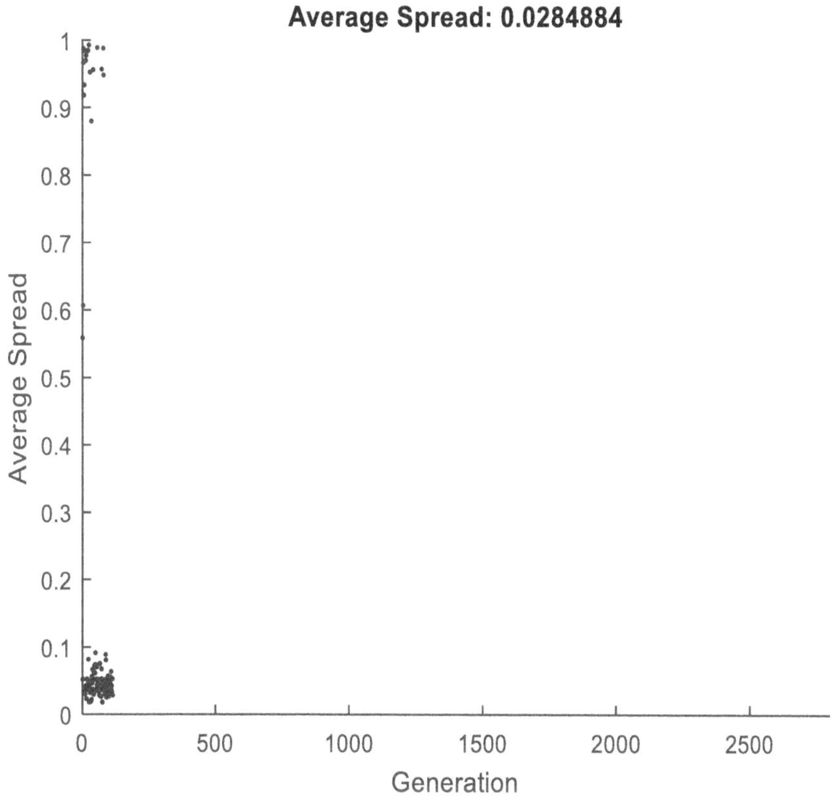


*Figure 11.3* Average spread with generation.

## 11.4 EXPERIMENTATION

The JYOTI computer numerical control VMC machine has been used for experimentation. A total of 25 tests have been carried out for validation of the results achieved by the mathematical model. The vertical milling center with the KISTLER dynamometer is shown in Figure 11.4. Here, AISI1020 material was used for experimentation with 130 BHN hardness and 7,860 $kg/m^3$ density. The chemical composition of AISI1020 is 0.565% of Mn, 0.026% of P, 0.039% of S, 0.197% of C, 99.100% of Fe. The work specimen before the experimentation with initial level, material properties testing level, and operation level are shown in Figure 11.5.

At the operation, the workpieces are located and held on the KISTLER dynamometer through the drilled hole.

During the experimentation total of 25 work specimens (100 x 50 x 10) has been prepared for 25 data sets. The work specimens after the end milling operation are shown in Figure 11.6.

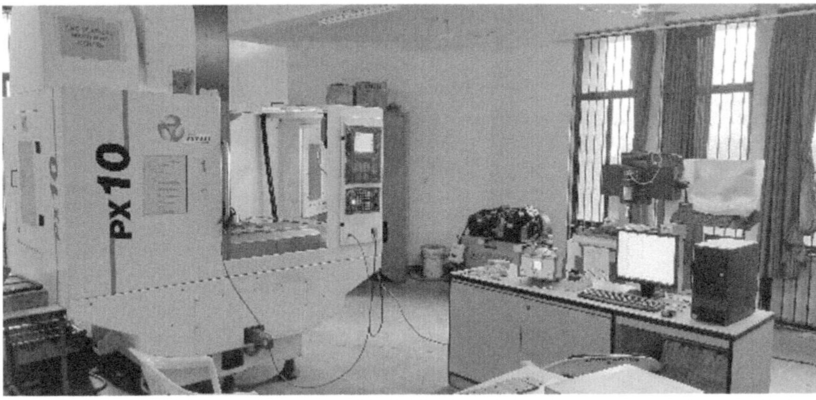


*Figure 11.4* VMC with KISTLER dynamometer.

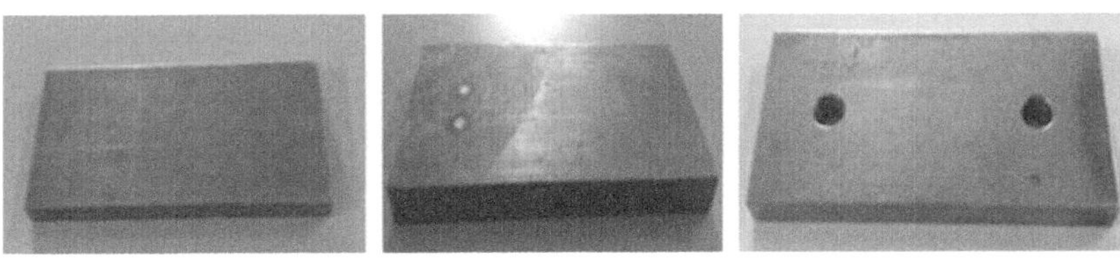

*Figure 11.5* Work specimen before the operation.

*Figure 11.6* Work specimen after the operation.

The 12 mm diameter end mill with 4 flutes has been used for cutting operation [16]. The experimental values of MRR calculated based on weight of workpieces before and after the experimentation. Tangential cutting force and torque have been measured through KISTLER dynamometer. Machining time has been measured through stopwatch and power has been calculated based on experimental values of tangential cutting force. The experimental results of 25 data sets are shown in Table A1 (Appendix).

## 11.5 RESULTS AND DISCUSSION

The comparison of results achieved by experiments and empirical equation-based mathematical model for machining time is shown in Figure 11.7. The outcomes indicate the mean square error between the numerical model and the test results of machining time has been seen as 8.35%.

The comparison of results achieved by experiments and empirical equation based mathematical model for MRR is shown in Figure 11.8. The outcomes indicate the mean square error between the numerical model and the test results of MRR has been seen as 7.97%.

The comparison of results achieved by experiments and empirical equation-based mathematical model for tangential cutting force is shown in Figure 11.9.

The outcomes indicate the mean square error between the numerical model and the test results of tangential cutting force has been seen as 7.57%. The comparison of results achieved by experiments and empirical equation-based mathematical model for torque is shown in Figure 11.10.

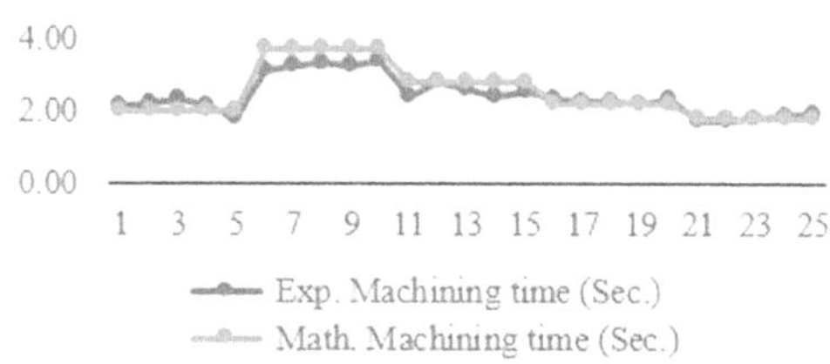


*Figure 11.7* Machining time results (math. model vs. experimental).

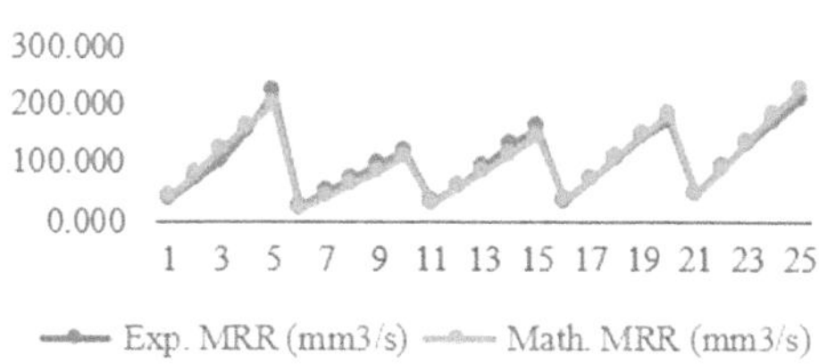


*Figure 11.8* MRR results (math. model vs. experimental).

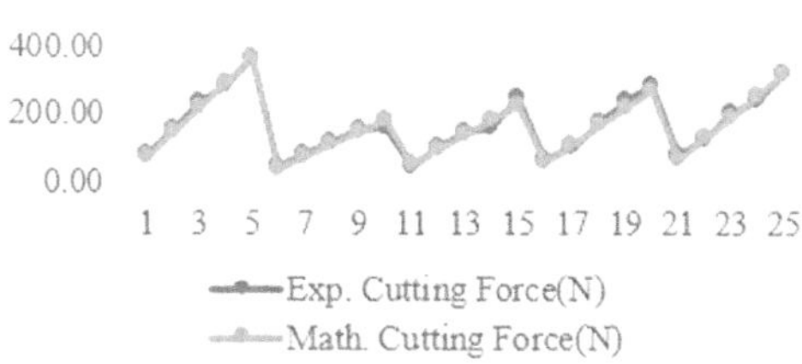


*Figure 11.9* Tangential cutting force results (math. model vs. experimental).

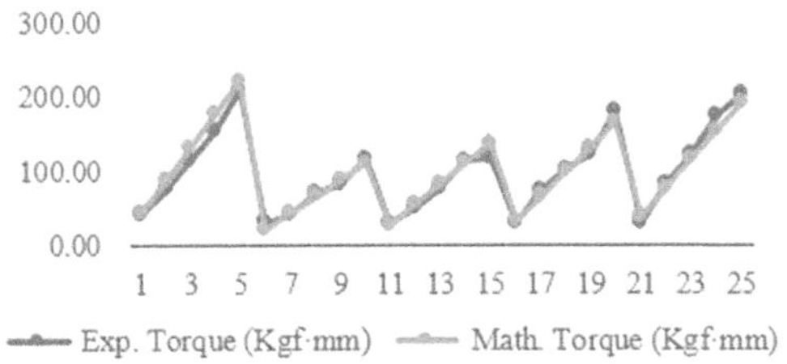


*Figure 11.10* Torque results (math. model vs. experimental).

The outcomes indicate the mean square error between the numerical model and the test results of torque have been seen as 10%.

The comparison of results achieved by experiments and empirical equation-based mathematical model for power is shown in Figure 11.11.

The outcomes indicate the mean square error between the numerical model and the test results of power have been seen as 7.57%. The comparison of results achieved by experiments and empirical equation-based mathematical model for tool life is shown in Figure 11.12. The outcomes indicate the mean square error between the numerical model and the test results of tool life have been seen as 0%.

It has been revealed that the results of the empirical-based mathematical model matched the experimental results. It shows positive agreement with experimental results. This empirical relation has been used in the MOGA tool for multi-objective optimization. The optimum combination of end milling process parameters of 18 non-dominated Pareto optimal solutions achieved through MOGA is shown in Table A2 (Appendix). The relation between the number of individuals and score is shown in Figure 11.13, which gives the range of minimum and maximum values of responses

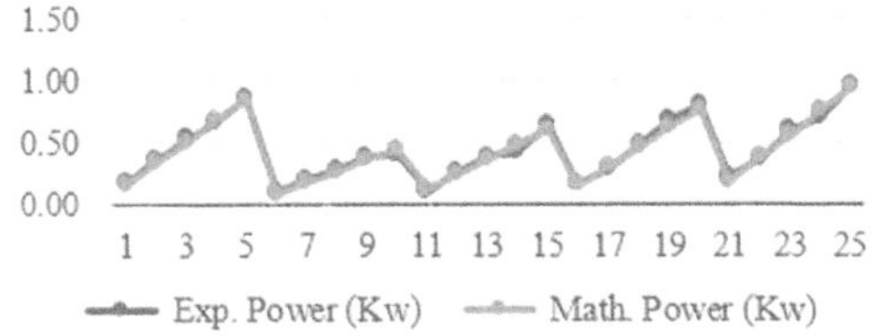


*Figure 11.11* Power results (math. model vs. experimental).

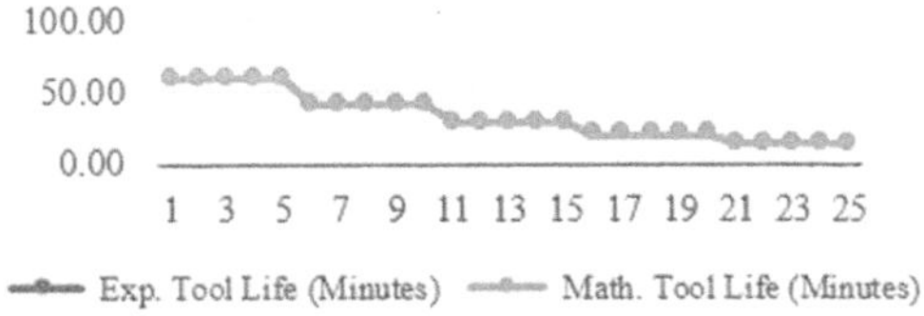


*Figure 11.12* Tool life results (math. model vs. experimental).

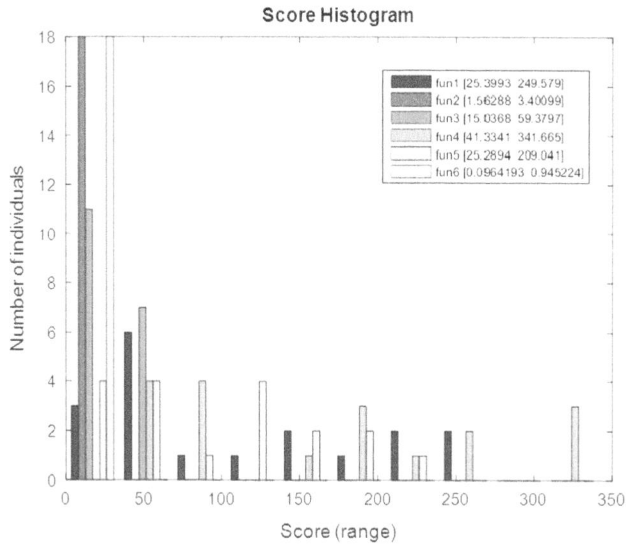


*Figure 11.13* Score histogram chart.

*Table 11.3* Responses with min. and max. values

| *Responses* | *Min. values* | *Max. values* |
|---|---|---|
| Machining time | 1.56288 | 3.40099 |
| MRR | 25.3993 | 249.59 |
| Tangential cutting force | 41.3341 | 341.665 |
| Torque | 25.2894 | 209.041 |
| Power | 0.0964193 | 0.945224 |
| Tool life | 15.0368 | 59.3797 |

related to constraint function. The minimum and maximum values of responses are shown in Table 11.3.

The point of the Pareto optimal front is shown in Figure 11.14 produced through the MOGA tool of selected responses.

One set of experiments has been conducted for validation of the MOGA results (Sr No. 10, Table A2). It has been found the percentage of error for machining time is 13.167%, MRR is 3.199%, tangential cutting force is 6.260%, torque is 0.527%, power is 6.260%, and tool life is 0.347%. Figure 11.15 indicated the good agreement of MOGA and experimental results. This multi-objective optimization approaches provide the optimum combination of input variables for given values of output parameters. It has been observed that all the solution generated by the MOGA tool is good. The preference of selection of Pareto optimal solutions depend upon the need of industries.

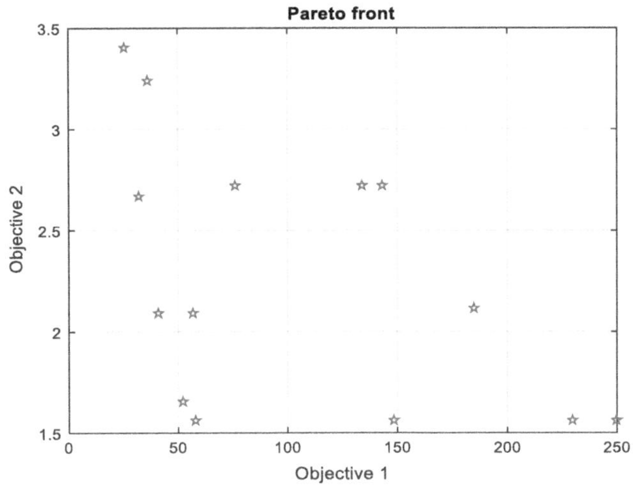


*Figure 11.14* Pareto optimal front chart.

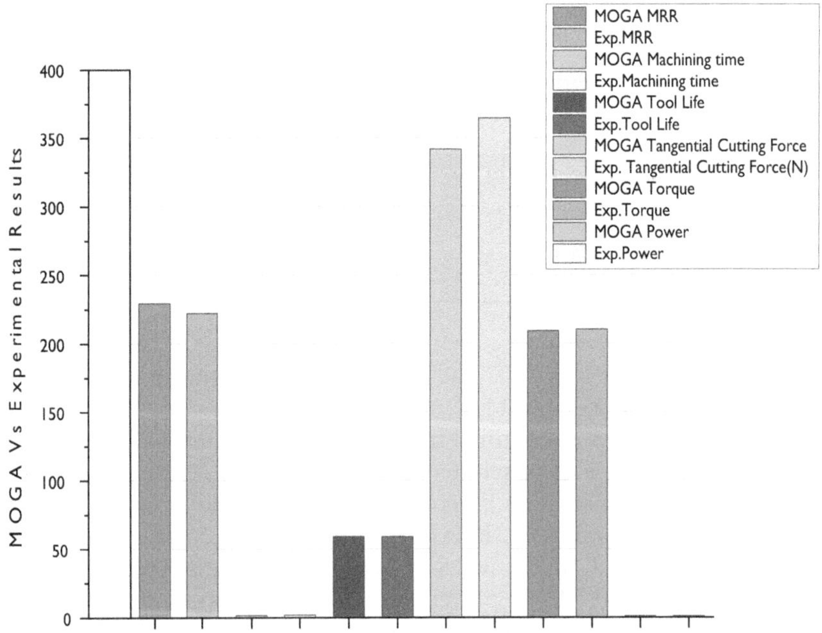


*Figure 11.15* Multi-objective genetic algorithm vs. experimental result.

## 11.6 CONCLUSION

The multi-objective genetic algorithm approaches utilized to multi-objective optimization based on empirical mathematical equations have been found very useful techniques in industries. The mathematical results provide average accuracy for machining time is 91.65%, material removal rate is 92.03%, tangential cutting force is 92.43%, torque is 90%, power is 92.43%, and tool life is 100%, when compared with results achieved by the experiment. MOGA gave the optimal solution with a maximum tool life is 59.343 minutes, MRR is 229.331 $m^3$/sec, minimum tangential cutting force is 341.665 N, torque is 209.041 Kgf.mm, power is 0.797, machining time is 1.563 sec with 140.016 m/min. of cutting speed, 0.24 mm/tooth of feed, and 0.950 mm of the depth of cut. It has been found the average accuracy of machining time is 86.833%, MRR is 96.801%, tangential cutting force is 93.74%, torque is 99.473%, power is 93.74%, and tool life is 99.653%. The results of the MOGA correspond with experimental results. This indicates that the multi-objective genetic algorithm is capable of giving the optimum cutting condition of input variables for the end milling operation. This study is useful for the industries to increase the productivity and decrease the consumption of power, cost, and time of production.

## REFERENCES

1. Sreenivasulu, R., C. S. Rao, and K. Ravindra. 2019. "Grey Based Taguchi Approach Integrated with Entropy Measurement for Optimization of Surface Roughness and Delamination Damage Factor during End Milling of GFRP Composites." *International Journal of Modern Manufacturing Technologies* 11 (2): 133–141.
2. Ajith Arul Daniel, S., R. Pugazhenthi, R. Kumar, and S. Vijayananth. 2019. "Multi Objective Prediction and Optimization of Control Parameters in the Milling of Aluminium Hybrid Metal Matrix Composites Using ANN and Taguchi -Grey Relational Analysis." *Defence Technology* 15 (4): 545–556. 10.1016/j.dt.2019.01.001.
3. Daniyan, I. A., I. Tlhabadira, K. Mpofu, and A. O. Adeodu. 2021. "Process Design and Optimization for the Milling Operation of Aluminum Alloy (AA6063 T6)." *Materials Today: Proceedings* 38 (1016): 536–543. 10.1016/j.matpr.2020.02.396.
4. Thepsonthi, T., and T. Özel. 2012. "Multi-Objective Process Optimization for Micro-End Milling of Ti-6Al-4V Titanium Alloy." *International Journal of Advanced Manufacturing Technology* 63 (9–12): 903–914. 10.1007/s00170-012-3980-z.
5. Nguyen, T. T., T. A. Nguyen, and Q. H. Trinh. 2020. "Optimization of Milling Parameters for Energy Savings and Surface Quality." *Arabian Journal for Science and Engineering* 45 (11): 9111–9125. 10.1007/s13369-020-04679-0.

6. Rajeswari, S., and P. S. Sivasakthivel. 2018. "Optimisation of Milling Parameters with Multi-Performance Characteristic on Al/SiC Metal Matrix Composite Using Grey-Fuzzy Logic Algorithm." *Multidiscipline Modeling in Materials and Structures* 14 (2): 284–305. 10.1108/MMMS-04-2017-0027.
7. Shaik, J. H., and J. Srinivas. 2020. "Optimal Design of Spindle-Tool System for Improving the Dynamic Stability in End-Milling Process." *Sadhana - Academy Proceedings in Engineering Sciences* 45 (1). 10.1007/s12046-020-1286-7.
8. Lusi, N., D. R. Pamuji, A. Fiveriati, A. Afandi, and G. S. Prayogo. 2020. "Application of Taguchi and Grey Relational Analysis for Parametric Optimization of End Milling Process of ASSAB-XW 42." 198 (Issat): 514–517. 10.2991/aer.k.201221.085.
9. Jebaraj, M., M. Pradeep Kumar, N. Yuvaraj, and G. M. Rahman. 2019. "Experimental Study of the Influence of the Process Parameters in the Milling of Al6082-T6 Alloy." *Materials and Manufacturing Processes* 34 (12): 1411–1427. 10.1080/10426914.2019.1594271.
10. Bhattacharya, A., 1984. "Metal Cutting Theory and Practice." Rev. and enl. ed edition Jamini Kanta Sen of Central Book Publishers.
11. Pradhan, M. K., M. Meena, S. Sen, and A. Singh. 2015. "Multi-Objective Optimization in End Milling of Al-6061 Using Taguchi Based G-PCA." *International Journal of Mechanical, Aerospace, Industrial, Mechatronic and Manufacturing Engineering* 9 (6): 1082–1088.
12. Tata McGRAW-HILL. 2000. "Production Technology." Seventeenth Reprint. hmt published.
13. Module 4 General Purpose Machine Tools, Version 2 ME, IIT Kharagpur.
14. Gajera, H. M., K. G. Dave, V. P. Darji, and K. Abhishek. 2019. "Optimization of Process Parameters of Direct Metal Laser Sintering Process Using Fuzzy-Based Desirability Function Approach." *Journal of the Brazilian Society of Mechanical Sciences and Engineering* 41 (3). 10.1007/s40430-019-1621-2.
15. PSG Design data book. 2000. Reprinted. PSG College of Technology.
16. Jain, R. K. 2001. "Production Technology." Khanna Publishers, Seventeenth edition.

# Appendix

*Table A1* Experimental results of 25 data set

| Sr no | Cutting speed in m/min | Feed rate in mm/ tooth | Depth of cut in mm | Hardness in BHN | Density in Kg/m$^3$ | Mach. time in Sec. | Material removal rate in mm$^3$/s | Tangential cutting Force in N | Torque in Kgf·mm | Power in Kw | Tool life in Min. |
|---|---|---|---|---|---|---|---|---|---|---|---|
| 1 | 140 | 0.24 | 0.2 | 130 | 7860 | 2.10 | 38.10 | 80.85 | 40.16 | 0.19 | 59.14 |
| 2 | 140 | 0.24 | 0.4 | 130 | 7860 | 2.19 | 73.06 | 151.00 | 76.59 | 0.35 | 59.14 |
| 3 | 140 | 0.24 | 0.6 | 130 | 7860 | 2.34 | 102.56 | 227.52 | 113.10 | 0.53 | 59.14 |
| 4 | 140 | 0.24 | 0.8 | 130 | 7860 | 2.10 | 152.38 | 279.91 | 153.89 | 0.65 | 59.14 |
| 5 | 140 | 0.24 | 1 | 130 | 7860 | 1.80 | 222.22 | 364.48 | 210.15 | 0.85 | 59.14 |
| 6 | 150 | 0.12 | 0.2 | 130 | 7860 | 3.10 | 25.81 | 38.26 | 31.37 | 0.10 | 40.32 |
| 7 | 150 | 0.12 | 0.4 | 130 | 7860 | 3.23 | 49.54 | 79.48 | 42.40 | 0.20 | 40.32 |
| 8 | 150 | 0.12 | 0.6 | 130 | 7860 | 3.31 | 72.51 | 112.87 | 70.02 | 0.28 | 40.32 |
| 9 | 150 | 0.12 | 0.8 | 130 | 7860 | 3.25 | 98.46 | 151.96 | 82.15 | 0.38 | 40.32 |
| 10 | 150 | 0.12 | 1 | 130 | 7860 | 3.39 | 117.99 | 159.22 | 116.19 | 0.40 | 40.32 |
| 11 | 160 | 0.15 | 0.2 | 130 | 7860 | 2.40 | 33.33 | 39.75 | 31.26 | 0.11 | 28.18 |
| 12 | 160 | 0.15 | 0.4 | 130 | 7860 | 2.76 | 57.97 | 101.41 | 51.39 | 0.27 | 28.18 |
| 13 | 160 | 0.15 | 0.6 | 130 | 7860 | 2.56 | 93.75 | 142.55 | 77.53 | 0.38 | 28.18 |
| 14 | 160 | 0.15 | 0.8 | 130 | 7860 | 2.40 | 133.33 | 160.45 | 113.10 | 0.43 | 28.18 |
| 15 | 160 | 0.15 | 1 | 130 | 7860 | 2.51 | 159.36 | 242.45 | 117.13 | 0.65 | 28.18 |
| 16 | 170 | 0.18 | 0.2 | 130 | 7860 | 2.30 | 34.78 | 60.63 | 29.79 | 0.17 | 20.13 |
| 17 | 170 | 0.18 | 0.4 | 130 | 7860 | 2.29 | 69.87 | 98.71 | 73.55 | 0.28 | 20.13 |
| 18 | 170 | 0.18 | 0.6 | 130 | 7860 | 2.23 | 107.62 | 170.48 | 102.11 | 0.48 | 20.13 |
| 19 | 170 | 0.18 | 0.8 | 130 | 7860 | 2.22 | 144.14 | 240.66 | 122.15 | 0.68 | 20.13 |

(Continued)

*Table A1* (continued)

| *Sr no* | *Cutting speed in m/min* | *Feed rate in mm/ tooth* | *Depth of cut in mm* | *Hardness in BHN* | *Density in Kg/m³* | *Mach. time in Sec.* | *Material removal rate in mm³/s* | *Tangential cutting Force in N* | *Torque in Kgf·mm* | *Power in Kw* | *Tool life in Min.* |
|---|---|---|---|---|---|---|---|---|---|---|---|
| 20 | 170 | 0.18 | 1 | 130 | 7860 | 2.33 | 171.67 | 281.52 | 179.73 | 0.80 | 20.13 |
| 21 | 180 | 0.21 | 0.2 | 130 | 7860 | 1.70 | 47.06 | 76.04 | 31.27 | 0.23 | 14.66 |
| 22 | 180 | 0.21 | 0.4 | 130 | 7860 | 1.75 | 91.43 | 120.22 | 82.05 | 0.36 | 14.66 |
| 23 | 180 | 0.21 | 0.6 | 130 | 7860 | 1.82 | 131.87 | 202.12 | 122.91 | 0.61 | 14.66 |
| 24 | 180 | 0.21 | 0.8 | 130 | 7860 | 1.86 | 172.04 | 235.60 | 174.58 | 0.71 | 14.66 |
| 25 | 180 | 0.21 | 1 | 130 | 7860 | 1.93 | 207.25 | 319.26 | 202.19 | 0.96 | 14.66 |

*Table A2* Non dominated pareto optimal solution

| Sr no | *Cutting speed in m/min* | *Feed rate in mm/ tooth* | *Depth of cut in mm* | *Hardness in BHN* | *Density in Kg/m³* | *Mach. time in Sec.* | *Material removal rate in mm³/s* | *Tangential cutting Force in N* | *Torque in Kgf·mm* | *Power in Kw* | *Tool life in Min.* |
|---|---|---|---|---|---|---|---|---|---|---|---|
| 1 | 178.38 | 0.138 | 0.2 | 120 | 7860 | 2.721 | 134.211 | 41.334 | 25.289 | 0.123 | 15.456 |
| 2 | 140 | 0.138 | 0.2 | 120 | 7860 | 2.72 | 76.089 | 41.334 | 25.289 | 0.096 | 59.38 |
| 3 | 167.188 | 0.139 | 0.92 | 120 | 7860 | 3.401 | 25.399 | 192.278 | 117.641 | 0.536 | 22.153 |
| 4 | 179.265 | 0.24 | 0.216 | 120 | 7860 | 1.564 | 57.771 | 77.549 | 47.447 | 0.232 | 15.037 |
| 5 | 179.265 | 0.227 | 0.92 | 120 | 7860 | 2.09 | 41.325 | 312.842 | 191.406 | 0.934 | 15.037 |
| 6 | 179.265 | 0.24 | 0.67 | 120 | 7860 | 1.563 | 249.579 | 240.979 | 147.438 | 0.72 | 15.037 |
| 7 | 155.439 | 0.178 | 0.204 | 120 | 7860 | 2.113 | 184.585 | 54.237 | 33.184 | 0.14 | 33.208 |
| 8 | 179.265 | 0.227 | 0.931 | 120 | 7860 | 1.654 | 52.228 | 316.455 | 193.617 | 0.945 | 15.037 |
| 9 | 155.439 | 0.146 | 0.95 | 120 | 7860 | 3.243 | 36.392 | 208.202 | 127.384 | 0.539 | 33.208 |
| **10** | **140.016** | **0.24** | **0.95** | **120** | **7860** | **1.563** | **229.331** | **341.665** | **209.041** | **0.797** | **59.343** |
| 11 | 149.204 | 0.178 | 0.763 | 120 | 7860 | 2.672 | 32.329 | 202.817 | 124.089 | 0.504 | 41.689 |
| 12 | 140.063 | 0.24 | 0.216 | 120 | 7860 | 1.564 | 148.134 | 77.549 | 47.447 | 0.181 | 59.233 |
| 13 | 149.204 | 0.24 | 0.204 | 120 | 7860 | 1.563 | 249.579 | 73.334 | 44.868 | 0.182 | 41.689 |
| 14 | 140 | 0.24 | 0.42 | 120 | 7860 | 1.563 | 229.331 | 151.067 | 92.428 | 0.352 | 59.38 |
| 15 | 140 | 0.138 | 0.2 | 120 | 7860 | 2.72 | 143.42 | 41.334 | 25.289 | 0.096 | 59.38 |
| 16 | 179.265 | 0.24 | 0.216 | 120 | 7860 | 1.564 | 57.771 | 77.549 | 47.447 | 0.232 | 15.037 |
| 17 | 179.265 | 0.227 | 0.795 | 120 | 7860 | 2.09 | 56.465 | 270.338 | 165.401 | 0.807 | 15.037 |
| 18 | 167.188 | 0.139 | 0.92 | 120 | 7860 | 3.401 | 25.399 | 192.278 | 117.641 | 0.536 | 22.153 |

Chapter 12

# Micro-Electrochemical Machining of Nimonic 263 Alloy: An Experimental Investigation and ANN-Based Prediction of Radial Over Cut

*Dilkush Bairwa*[1], *Dr Ravi Pratap Singh*[1], *Dr Ravinder Kataria*[2], *Dr Sandeep Singhal*[3], *Dr Narendra Kumar*[1], *Shailendra Chauhan*[1], *and Madhusudan Painuly*[1]

[1]Department of Industrial and Production Engineering, Dr B R Ambedkar National Institute of Technology, Jalandhar, Punjab, India
[2]National Institute of Fashion Technology, Srinagar, Jammu & Kashmir, India
[3]Department of Mechanical Engineering, National Institute of Technology, Kurukshetra, Haryana, India

**CONTENTS**



## 12.1 INTRODUCTION

In micro-electrochemical machining, many physical and chemical processes happen at the same time, such as heat transfer, electrolyte flow, electrochemical reactions, and the distribution of electric fields, etc. Because it is contactless, micro-electrochemical machining (micro-ECM) has the

DOI: 10.1201/9781003258421-13

potential to provide high-quality surface integrity. The ECM method is based on Faraday's law of electrolysis. During the electrochemical machining process ($NaNO_3$), both the tool electrode and the workpiece are submerged in an electrolytic concentration of sodium nitrate ($NaNO_3$) [1]. We have sustained a tiny gap between the two electrodes with no tool wear, mechanical forces, or residual stresses by applying a consistent magnitude of potential between the tool electrode and the workpiece. The workpiece behaves as an anode (positive) and the tool electrode behaves as a cathode (negative) in electrochemical machining (ECM), and DC power is delivered between the tool electrode and the workpiece, enabling us to dissolve the anodic substance.

### 12.1.1 Micro-Electrochemical Machining

Micro-electrochemical machining is used in a variety of industrial areas, including aerospace engineering, automotive engineering, and microelectronics. Micro-electrochemical machining would be used to create three-dimensional micro-structures such as micro-holes and micro-grooves. Figure 12.1

In micro-electrochemical machining, the workpiece is secured in a frame and the tool is introduced into a tool post, and both are kept extremely close to each other so that the inter electrode gap (IEG) may be maintained. The inter electrode gap is the space between the tool and the workpiece. Following immersion of the tool and workpiece in an electrolyte solution (such as NaCl), a voltage difference across the tool and workpiece is applied, and material removal commences.

### 12.1.2 Materials

The Nimonic-263 alloy material was utilised in this experiment because it has strong corrosion resistance, high tensile strength, high temperature

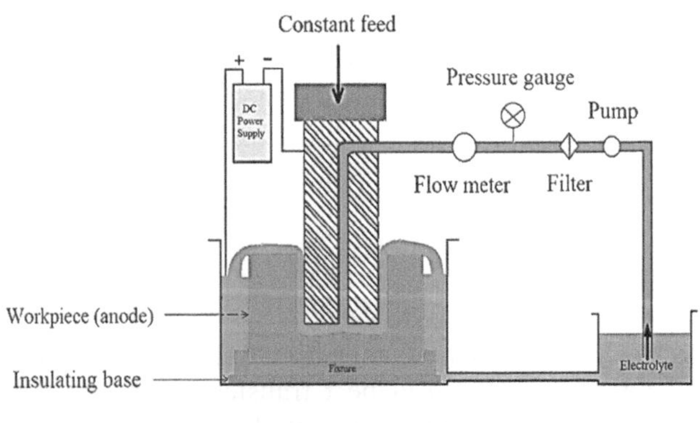


*Figure 12.1* Schematic diagram of ECM.

resistance, and high yield strength. The Nomonic-263 alloy material has been utilised in the aerospace industry in aviation turbine engines for casings, rings, and fabrication sheets. Nimonic-263 alloy has a high heat conductivity. It has a high thermal ductility in welded constructions.

Shunda Zhan et al. (2020) [2] The machining precision of the blade leading edge was enhanced by 41.79% by applying several machining parameters such as inlet/outlet pressure (0.30/0.08 MPa), applied voltage (20 V), feed speed (0.3 mm/min), and initial gap (0.5 mm). Feng Wang and colleagues.2020 [3] The surface quality of a rhombus hole could have been enhanced by using the synchronized electrochemical machining technique, and the radius of fillet and slope of sidewall were essentially lowered by raising the amplitude, which may increase the accuracy of the rhombus hole. With a feed rate of 0.4 mm/min, the surface quality has increased. Thomas Bergs and colleagues.2020 [4] Conducted ECM is performed on pure cementite material (Fe3c) using a tool material of 42CrMo4 that is rotated at a speed of 20 to 40 m/s. In this research, two important machining factors (such as feed rate Vf and voltage U) were investigated, and the conclusion of an electrochemical machining process certificate was successfully completed for the modification of the localised phase composition and surface grinding. Wenjian Cao and colleagues.2020 [5] Performed counter-rotating electrochemical machining (CRECM) to machine thin wall going to revolve parts, and the initial inter electrode gap decreased (0.272–0.158 mm) with increasing voltage (11–15 V), but continued to increase after transitional period because comparative equilibrium MRR is greater than the corresponding feed rate. Guodong Liu.2020 [6] The group of researchers led by using sidewall insulating coatings on the electrode, it is possible to prevent stray corrosion and the sidewall slope of the machined structure has been lowered to 5.5 degrees by using nanofabrication photomasks to design all electrode forms and therefore influence dimension accuracy. Chen, X.L., and colleagues .2020 [7] A 20% duty cycle for jet flow–type pulses was shown to be more useful in reducing microgroove surface roughness when compared to other electrolyte flow patterns. Slit nozzle and standard deviation were operating with parameters such as temperature (250°C), frequency (2 kHz), duty cycle (20%, 40%, 60%, and 80%) of the machining time (10 s), voltage (20, 15, 20, 25 V), and pressure (250 psi) (0.4 MPa).

## 12.2 MATERIALS AND METHODS

In order to carry out experiments on Nimonic-263 micro-ECM alloy, workpieces made of Nimonic-263 alloy of 1 cm x 1 cm have been employed in the experimental work. For micro-drilling, micro-electrochemical machining was utilised. Experiments have been conducted to study the impacts of various machining inputs, such as the applied voltage, feed rate, pulse

*Table 12.1* Chemical composition of Nimonic-263 alloy

| *Sr. No.* | *Elements* | *Content (%)* |
|---|---|---|
| 1 | Nickel, Ni | 49 |
| 2 | Cobalt, Co | 19–21 |
| 3 | Chromium, Cr | 19–21 |
| 4 | Molybdenum, Mo | 5.60–6.10 |
| 5 | Titanium, Ti | 1.90–2.40 |
| 6 | Manganese, Mn | 0.60 |

frequency, and electrolyte solution in order to generate micro-holes with a high-accuracy profile and acceptable surface quality. Table 12.1 displays the material's elemental constituents. It was used for investigations in micro-electrochemical machining of nickel-based super alloy Nimonic-263. The Taguchi L16 orthogonal array method [8] was used to design the tests. There were four present levels for four different parameters. A micro-ECM method with the best-chosen parameters was used to make initial holes on Nimonic-263 alloy, which were subsequently machined independently utilising micro-ECM setups. After all of the tests, the best parameters for micro-drilling have been identified. Sodium nitrate ($NaNO_3$) was the electrolyte solution employed in this experiment, while copper was used as the tool electrode. Maximum voltage is 10 V, pulse frequencies are 5 KHz to 150 KHz, and feed rates are 20–50/min. In order to conduct additional research, the average answer value has been obtained. There was a lot of attention paid to this experimental metric, the radial over cut (ROC). Taguchi's L16 orthogonal array approach was utilised to examine the impact of machining input parameters on response parameters like radial over cut in this experiment.

It was necessary to choose inputs such as the voltage used, the pulse frequency, the electrolyte solution and the feed rate [8]. Four levels of factor classification are available to users. For several experiments, the electrolyte solution was regularly changed and refilled with new electrolyte solution. Table 12.2 lists the machine inputs that were chosen, along with the levels that were assigned to each one. The micro-holes were made using a systematic approach to compare productivity and accuracy, and significant

*Table 12.2* Process parameters with their selected levels of micro-ECM

| *Parameters* | *Level 1* | *Level 2* | *Level 3* | *Level 4* |
|---|---|---|---|---|
| Applied voltage (V) | 7 | 8 | 9 | 10 |
| Feed rate (mm/min) | 0.02 | 0.03 | 0.04 | 0.05 |
| Electrolyte concentration (g/l) | 40 | 50 | 60 | 70 |
| Pulse frequency (KHz) | 50 | 75 | 100 | 125 |

*Table 12.3* L16 orthogonal array of micro-ECM with average output response (avg ROC)

| *Exp. no* | *AV (volt)* | *EC (g/l)* | *PF (KHz)* | *FR (mm/min)* | *Average radial over cut (micro-meter)* |
|---|---|---|---|---|---|
| 1. | 7 | 40 | 50 | 0.02 | 164.8635 |
| 2. | 7 | 50 | 75 | 0.03 | 182.3283 |
| 3. | 7 | 60 | 100 | 0.04 | 236.0387 |
| 4. | 7 | 70 | 125 | 0.05 | 264.6620 |
| 5. | 8 | 40 | 75 | 0.04 | 217.7725 |
| 6. | 8 | 50 | 50 | 0.05 | 291.6243 |
| 7. | 8 | 60 | 125 | 0.02 | 226.9663 |
| 8. | 8 | 70 | 100 | 0.03 | 176.5900 |
| 9. | 9 | 40 | 100 | 0.05 | 208.3970 |
| 10. | 9 | 50 | 125 | 0.04 | 220.6253 |
| 11. | 9 | 60 | 50 | 0.03 | 166.8317 |
| 12. | 9 | 70 | 75 | 0.02 | 254.1427 |
| 13. | 10 | 40 | 125 | 0.03 | 166.8257 |
| 14. | 10 | 50 | 100 | 0.02 | 233.9960 |
| 15. | 10 | 60 | 75 | 0.05 | 286.9577 |
| 16. | 10 | 70 | 50 | 0.04 | 309.0561 |

reaction was quantified as radial over cut. There are 16 orthogonal arrays (OA) in this experiment, which are shown in Table 12.3.

## 12.3 RESULTS AND DISCUSSION

As an output parameter is affected by the parameter level, the signal-to-noise (S/N) ratio has been used to assess this impact. Using a sequence of 16 tests, micro-ECM may be used to drill micro-holes.

Table 12.3 shows the average radial overcut and Table 12.4 shows the S/N ratio of each output response. The best value for the factor is the one

*Table 12.4* ANOVA results for ROC (SN ratios)

| *Source* | *DF* | *Seq SS* | *Adj SS* | *Adj MS* | *F* | *P* |
|---|---|---|---|---|---|---|
| AV (volt) | 3 | 4.547 | 4.547 | 1.5157 | 1.98 | 0.295 |
| EC (g/l) | 3 | 11.871 | 11.871 | 3.9569 | 5.16 | 0.046 |
| PF (KHz) | 3 | 1.338 | 1.338 | 0.4458 | 0.58 | 0.667 |
| FR (mm/min) | 3 | 29.199 | 29.199 | 9.7331 | 12.68 | 0.033 |
| Residual error | 3 | 2.302 | 2.302 | 0.7673 | | |
| Total | 15 | 49.257 | | | | |

*Table 12.5* ANOVA results for ROC (means)

| *Source* | *DF* | *Seq SS* | *Adj SS* | *Adj MS* | *F* | *P* |
|---|---|---|---|---|---|---|
| AV (volt) | 3 | 3686 | 3686 | 1228.8 | 2.48 | 0.238 |
| EC (g/l) | 3 | 8049 | 8049 | 2683.2 | 5.42 | 0.039 |
| PF (KHz) | 3 | 1298 | 1298 | 432.6 | 0.87 | 0.543 |
| FR (mm/min) | 3 | 18344 | 18344 | 6114.8 | 12.35 | 0.034 |
| Residual error | 3 | 1486 | 1486 | 495.3 | | |
| Total | 15 | 32864 | | | | |

with the highest mean S/N ratio [8]. The ANOVA findings for ROC for SN ratios are shown in Table 12.4 (below). Data from the ROC (radial over cut) ANOVA is shown in Table 12.5.

## 12.3.1 Radial Over Cut for Means

The process machining effiency is evaluated by this response. Figure 12.2 shows the primary effect chart: (applied voltage 7 V; electrolyte concentration 40 g/l; pulse frequency 100 KHz; feed rate 0.03 mm per minute) have been selected as the optimal input parameters. A1, B1, C3, and D2 are examples of the input parameters.

Observing the radial over cut values, this graph shows the primary impacts plot for means.

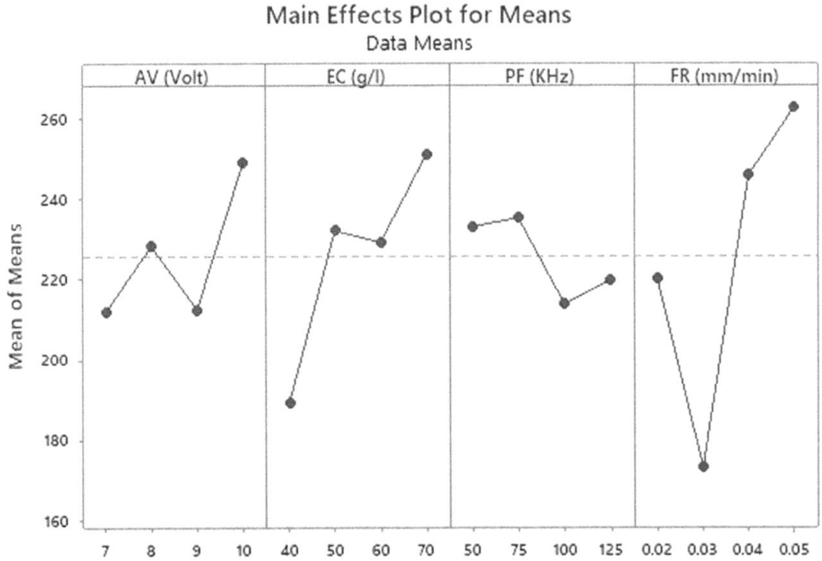


*Figure 12.2* Main effects plots for radial over cut (means).

- The radial over cut values gradually climbed from 7 V to 8 V in the beginning; however, after 8 V the values significantly reduced. This is followed by a further 10 V rise in radial over cut. First, a minimal radial overcut is performed at a voltage of 7 V.
- When the electrolyte concentration was raised from 40 g/l to 50 g/l, the radial over cut increased gradually, before sharply decreasing till 60 g/l and then slightly rising.
- Radial over cut increases sharply from pulse frequency of 100 KHz to 125 KHz, which is the lowest at which radial overcut is allowed.
- Radial over cut is reduced by 0.03 mm/min feed rate at the beginning of the graph. After then, the radial overcut is somewhat increased.

### 12.3.2 Radial Over Cut for SN Ratios

The radial overcut values are shown in this graph, which shows the major impacts plot for SN ratios. Figure 12.3

### 12.3.3 Residual Plots for Means

- There are radial over cut values at various factors, as seen in this graph of residual plots for means ratios.
- Figure 12.4 shows the residual normal probability plots for radial over cuts (Figure 12.4). The normal probability curve is shown as a single

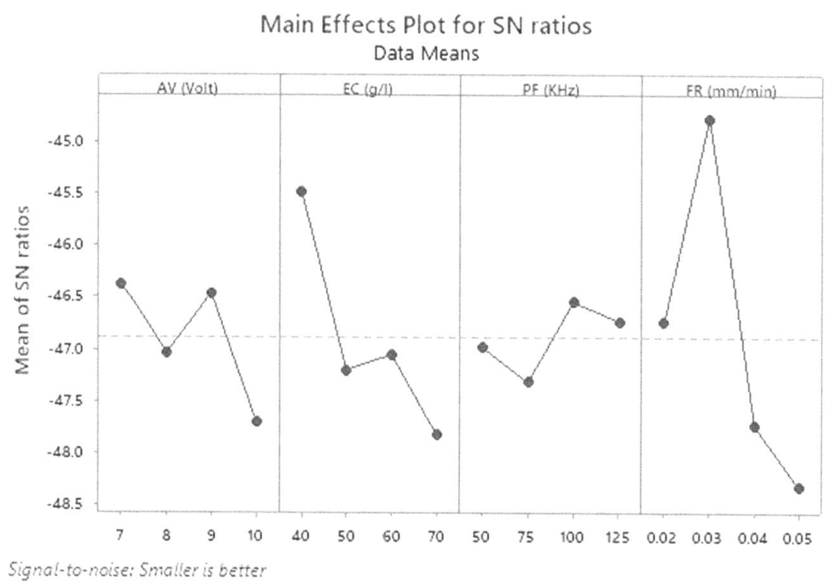


*Figure 12.3* Main effects plots for radial over cut (SN ratios).

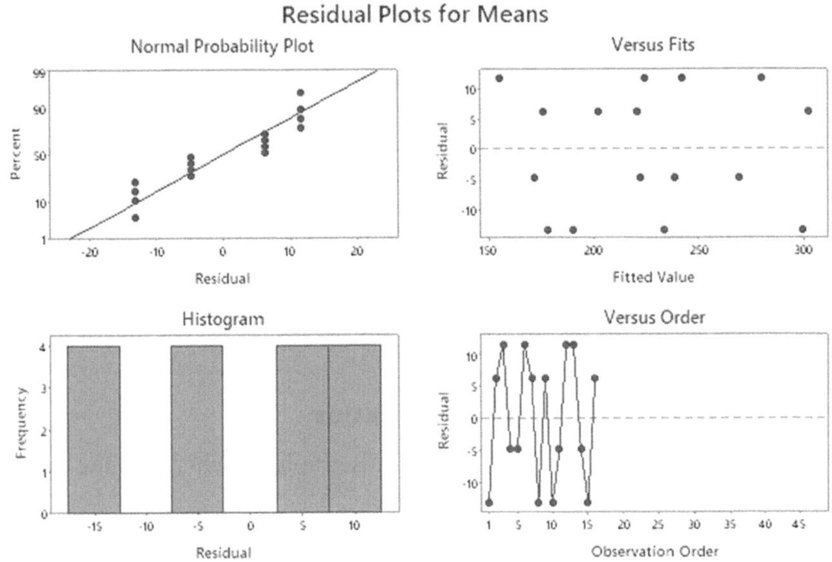


*Figure 12.4* Residual plots for radial over cut (means).

direction on a graph. As shown by normal probability plots displaying the polynomial regression model, all data points are located within a few hundredths of a degree of the polynomial regression line.

- Normal probability plots depict the radial overcut values between the typical percentage of probability and the residuals.
- The radial over cut values between residuals and fitted values graphs are shown in the residuals vs. fits graph. A "residuals vs. fits plot" is the most used visual in a residual analysis. Residuals and fitted values are shown in this image, which shows non-linearity, mismatched error variances, and outliers.
- In order to describe the radial over cut values, a histogram depicts the relationship between frequency and residuals. It is possible to utilise the residual histogram to determine whether or not the variance is uniformly distributed.
- Inverse order plots show the relationship between residuals and the sequence in which observations were collected. The residual vs. observation order graph came into play to identify the mistake in the sequence that is near to each other, regardless of how the data were gathered in space.

### 12.3.4 Residual Plots for SN Ratios

There are radial over cut values at various factors, as seen in this graph, which shows the residual plots for SN ratios. The residual plot for SN ratios

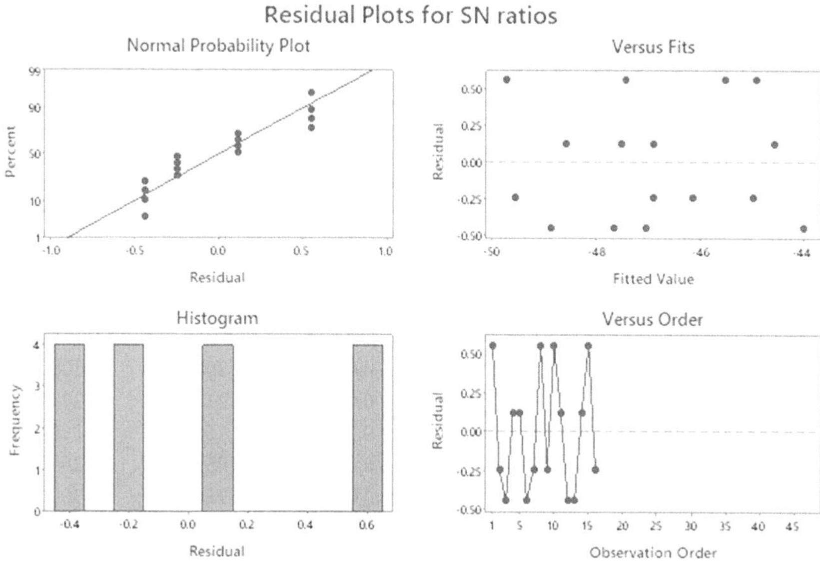


*Figure 12.5* Residual plots for radial over cut (SN ratios).

is almost identical to the residual plot for means in that it represents all values. Figure 12.5

## 12.4 PREDICTION OF RADIAL OVER CUT (ROC) VALUES USING AN ARTIFICIAL NEURAL NETWORK

Based on the input parameters such as feed rate, electrolyte concentration, pulse frequency, and applied voltage, the radial over cut has been calculated at the time of machining of Nimonic-263 alloy. A multi-layer neural network where two hidden layers were selected of 20 neurons each is added to predict the data. The ANN model has been trained by the Bayesian regulation of MATLAB. This regularization can handle improper data by solving the problems underfitting and overfitting. Table 12.6 shows the statistical data of radial over cut for ANN.

### 12.4.1 Artificial Neural Network Model

The artificial neural network used is shown in Figure 12.6: four input neurons viz feed rate, electrolyte concentration, pulse frequency, and applied voltage; also two hidden layers contain 20 neurons each (as per accuracy level) and finally, there is one output neurons (radial over cut). For training the data Levenberg-Marquardt algorithm was used for better

*Table 12.6* Statistical data of radial over cut for ANN

| *Exp. no.* | *AV (volt)* | *EC (g/l)* | *PF (KHz)* | *FR (mm/min)* | *Avg OC (mean)* | *ANN OC* |
|---|---|---|---|---|---|---|
| 1 | 7 | 40 | 50 | 0.02 | 164.8635 | 164.8636 |
| 2 | 7 | 50 | 75 | 0.03 | 182.3283 | 165.6313 |
| 3 | 7 | 60 | 100 | 0.04 | 236.0387 | 226.9868 |
| 4 | 7 | 70 | 125 | 0.05 | 264.6620 | 264.662 |
| 5 | 8 | 40 | 75 | 0.04 | 217.7725 | 217.7725 |
| 6 | 8 | 50 | 50 | 0.05 | 291.6243 | 291.6243 |
| 7 | 8 | 60 | 125 | 0.02 | 226.9663 | 233.0124 |
| 8 | 8 | 70 | 100 | 0.03 | 176.5900 | 176.59 |
| 9 | 9 | 40 | 100 | 0.05 | 208.3970 | 208.397 |
| 10 | 9 | 50 | 125 | 0.04 | 220.6253 | 220.6253 |
| 11 | 9 | 60 | 50 | 0.03 | 166.8317 | 166.8317 |
| 12 | 9 | 70 | 75 | 0.02 | 254.1427 | 254.1427 |
| 13 | 10 | 40 | 125 | 0.03 | 166.8257 | 300.89 |
| 14 | 10 | 50 | 100 | 0.02 | 233.9960 | 233.996 |
| 15 | 10 | 60 | 75 | 0.05 | 286.9577 | 286.9577 |
| 16 | 10 | 70 | 50 | 0.04 | 309.0561 | 309.056 |

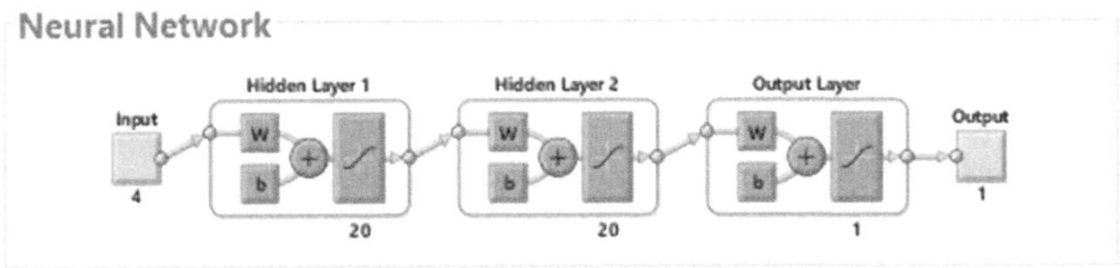


*Figure 12.6* Artificial neural network structure in the model.

results, further random data division, and for performance mean square error are used as parameters for ANN. According to the Figure 12.6, w and b are the assigned weights and biases for the neurons; further, the value of the radial over cut has been predicted by using a 4–20-20-1 structure for cryogenic-treated CBN due to better accuracy. Figure 12.6 shows the structure used in the model.

Figure 12.7 reveals that the experimental runs 2, 3, 7, and 13 have different values compared to the predicted value of ANN. Further, experiment numbers 4, 6, 13, and 16 have higher radial over cut values. Figure 12.7 shows the comparison between predicted values through ANN and experimental values for NC-CBN.

Figure 12.8 shows a plot between mean square error (MSE) and six epochs non-cryogenic-treated CBN data set. In ANN, an epoch is one cycle

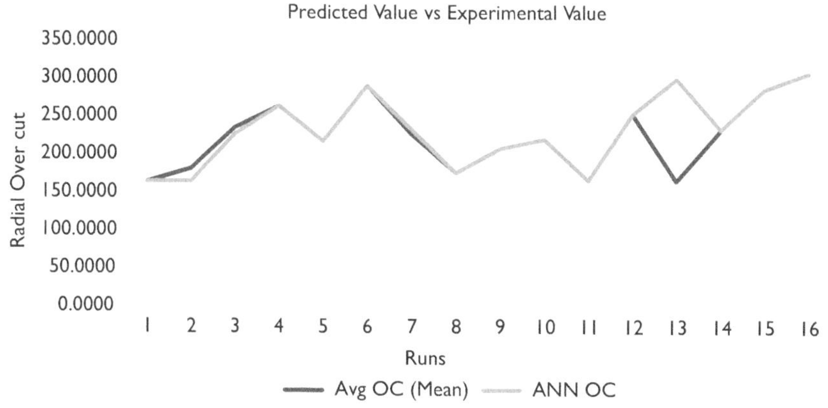


*Figure 12.7* Comparison of predicted and experimental values for ROC.

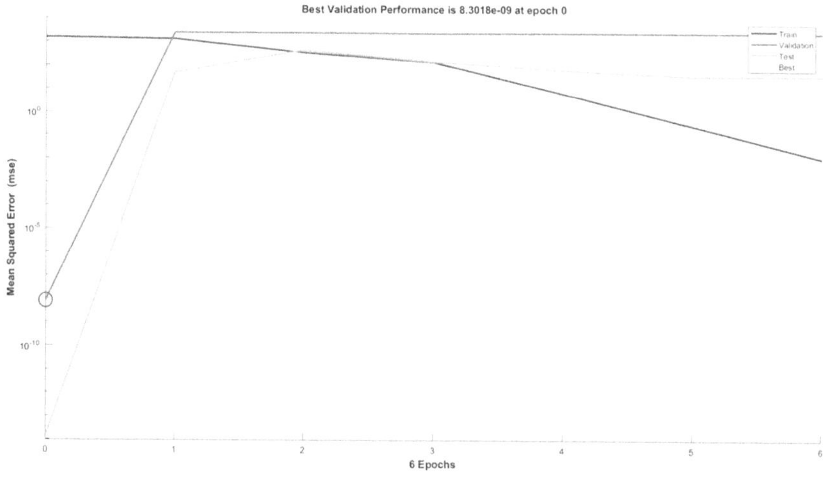


*Figure 12.8* Iteration graph for ROC through ANN.

through the full training datas et. Figure 12.8 further shows that the best validation has occurred at one epoch, which means one cycle is implemented for full training of the data set in the case of non-cryogenic-treated CBN. Figure 12.8 reveals the case of a non-cryogenic-treated CBN when predicted by the ANN model of the experimental values.

Figure 12.9 shows the comparison between predicted values through ANN and experimental values for cryogenic-CBN. Figure 12.9 shows the plot between target value and output value (predicted values).

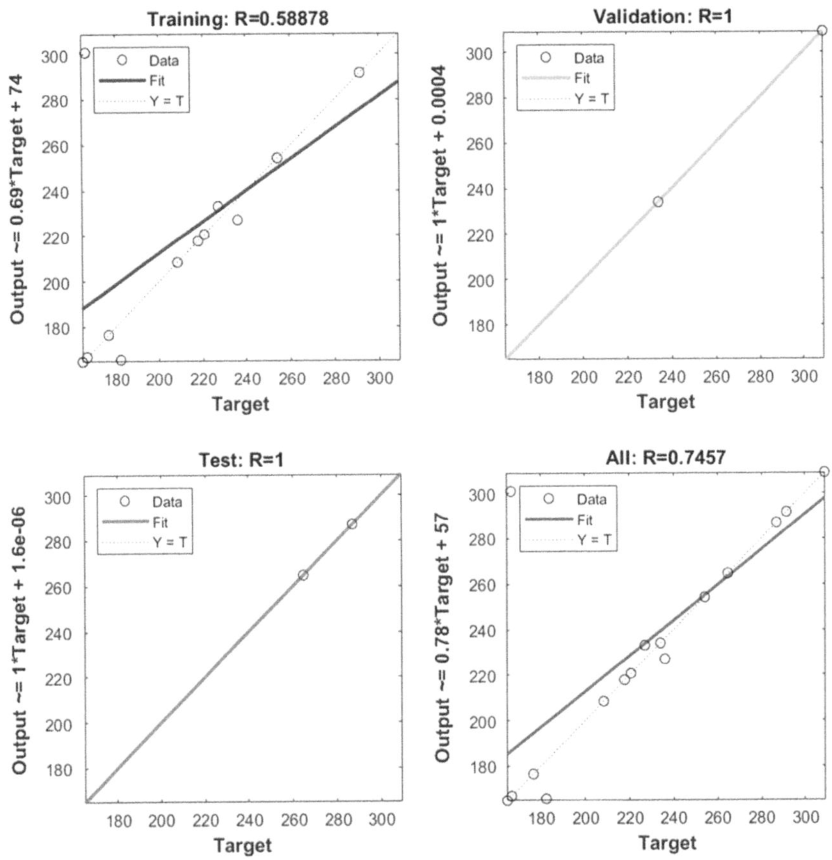


*Figure 12.9* Regression plots through ANN of NC-CB for ROC.

## 12.5 CONCLUSION

As part of this research, the radial over cut was investigated using an experimental technique including machining input parameters, which included micro-drilling operations to create micro-holes in the Nimonic-263 alloy. Using a methodical approach, all of the data needed to plan and conduct tests was collected. The following findings are based on data from four stages of experimentation. Use of the Taguchi L16 orthogonal approach and artificial neural network model may be used to calculate the optimal process parameters and the expected data of a radial over cut and also detect the error between experimental values and anticipated values of a radial over cut.

- The optimal parametric observations for meanings A1, B1, C3, and D2 (applied voltage 7 V, electrolyte concentration 40 g/l, pulse frequency

100 KHz, and feed rate 0.03 mm/min) have been explored based on micro-ECM operations.

- For SN ratios A1, B1, C3, and D2, the optimal parametric observation is based on the micro-ECM operation.
- Applying voltage and feed rate are two of the most important characteristics that affect the workpiece.
- With these ideal conditions (applied voltage 7 V, electrolyte concentration 40 g/l, pulse frequency 100 KHz, and feed rate 0.03 mm/min), the least quantity of material is eliminated. The MINITAB programme has produced better outcomes in terms of the DOE approach.
- Artificial neural networks (ANNs) have been used to predict radial over cut values.
- Because of its superior accuracy, the 4–20-20-1 structure has been used to forecast the radial over cut value for both cryogenically treated CBN samples.

## REFERENCES


1. Y. Wang, Z. Xu, and A. Zhang, "Anodic characteristics and electrochemical machining of two typical γ-TiAl alloys and its quantitative dissolution model in NaNO3 solution," *Electrochim. Acta*, vol. 331, p. 135429, 2020, doi: 10.1016/j.electacta.2019.135429
2. S. Zhan, and Y. Zhao, "Plasma-assisted electrochemical machining of microtools and microstructures," *Int. J. Mach. Tools Manuf.*, vol. 156, no. 1088, p. 103596, 2020, doi: 10.1016/j.ijmachtools.2020.103596
3. F. Wang, J. Yao, and M. Kang, "Electrochemical machining of a rhombus hole with synchronization of pulse current and low-frequency oscillations," *J. Manuf. Process.*, vol. 57, no. 40, pp. 91–104, 2020, doi: 10.1016/j.jmapro.2020.06.014
4. T. Bergs, and S. Harst, "Development of a process signature for electrochemical machining," *CIRP Ann.*, vol. 69, no. 1, pp. 153–156, 2020, doi: 10.1016/j.cirp.2020.04.078
5. W. Cao, D. Wang, and D. Zhu, "Modeling and experimental validation of interelectrode gap in counter-rotating electrochemical machining," *Int. J. Mech. Sci.*, vol. 187, no. June, p. 105920, 2020, doi: 10.1016/j.ijmecsci.2020.105920
6. G. Liu, H. Tong, Y. Li, and H. Zhong, "Novel structure of a sidewall-insulated hollow electrode for micro electrochemical machining," *Precis. Eng.*, vol. 72, no. April, pp. 356–369, 2021, doi: 10.1016/j.precisioneng.2021.05.009
7. X. L. Chen *et al.*, "Investigation on the electrochemical machining of micro groove using masked porous cathode," *J. Mater. Process. Technol.*, vol. 276, no. May 2019, 2020, doi: 10.1016/j.jmatprotec.2019.116406
8. P. V. Sivaprasad, K. Panneerselvam, and A. N. Haq, "A comparative assessment in sequential μ-drilling of Hastelloy-X using laser in combination with μ-EDM and μ-ECM," *J. Brazilian Soc. Mech. Sci. Eng.*, vol. 43, no. 7, pp. 1–19, 2021, doi: 10.1007/s40430-021-03068-4

# Index